ISAAC ASIMOV'S BOOK OF SCIENCE AND NATURE QUOTATIONS

Isaac Asimov's BOOK OF SCIENCE AND NATURE QUOTATIONS

Edited by

Isaac Asimov
and Jason A. Shulman

A BLUE CLIFF EDITIONS BOOK

Weidenfeld & Nicolson
New York

Published by Weidenfeld & Nicolson, New York
A Division of Wheatland Corporation
10 East 53rd Street
New York, NY 10022

Published in Canada by General Publishing Company, Ltd.

Due to limitations of space, permissions appear on page 337.

Library of Congress Cataloging-in-Publication Data

Isaac Asimov's book of science and nature quotations / edited by Isaac Asimov and Jason A. Shulman.—1st ed.
p. cm.
"A Blue Cliff editions book."
Includes index.
ISBN 1-555-84111-2
1. Science—Quotations, maxims, etc. I. Asimov, Isaac, 1920– II. Shulman, Jason. III. Title: Book of science and nature quotations.
Q173.I83 1988 87-22489
500—dc19 CIP

Manufactured in the United States of America
Designed by Irving Perkins Associates, Inc.
First Edition
10 9 8 7 6 5 4 3 2 1

FOR MY DAUGHTER ARIANA ELIZABETH:

May the Seas stay Deep
and the Sky stay High
May your Heart stay Open
and Beauty fill your Eye.

J. A. S.

Acknowledgments

THIS BOOK could not have been created without a number of people I would like to acknowledge here:

Amy Appleby, as head researcher, was responsible for gathering most of the raw material used in the book. Her skill in finding quotations for every subject I could conceive of (and some she invented!) was truly amazing. This project could not have gotten out of its infancy without her care and work.

Susan Walker not only added many of the quotations, but did much of the original research in finding quotations from news stories and scientific abstracts alike. Her contribution has been to add primary material to this book, and thereby make it useful to both scholars and lay people for years to come.

Elizabeth Guss added additional primary material and helped us add the spice that made the final work more flavorful.

Peter Pickow created the original programming we used to collect the database that eventually became this book. We are grateful to have a Pickow original in our computer.

Mark Polizzotti, our editor at Weidenfeld & Nicolson, not only read and commented in detail on the entire manuscript, but like a good gardener helped us pull out weeds and add new plants in barren spots.

Jackie Ogburn, as managing editor of Blue Cliff Editions, had the truly mammoth job of organizing all the raw material into a recognizably literate form. This took erudition, intelligence, fortitude, and patience. It took the ability to juggle thousands of bits of information on the computer and have it turn out right. It took a knowledge of philosophy, science, and the art of bookmaking. Ms. Ogburn also played a vital role in the basic organization of this book, from adding quotations in areas of dearth and implementing changes in categories that enhanced the final structure of the work, to adding her editorial voice in the final choice of material itself.

Finally, my thanks to Isaac Asimov, the inspiration of my boyhood. He is the man who, more than any other, taught me that the subject of science can be passionate and exciting, and that the study of science warms the soul while it enlightens the mind.

Thank you to all.
JASON SHULMAN

Contents

Foreword

BY ISAAC ASIMOV

THERE ARE five billion people on Earth, and I should guess that at any given moment one or two billion of us are speaking. And I should further guess that virtually nothing anyone says is memorable. A statement may give useful information, such as, "It's five o'clock," or it may be very precious to someone, as with, "I love you." But these are ephemeral things.

There are times, however, when someone turns a phrase that seems so clever or so apt or so enlightening or so true, that the statement flies from person to person and gives pleasure at each passage. The statement becomes a "quotation."

Quotations do not necessarily have extended lifetimes. Some are appropriate only to the immediate occasion and lose their force with time. Some are never written down and, if too much time passes, they may be forgotten. Some are too lengthy and complex, or too weighty, to remain long in our minds even if they *are* written down.

A quotation has the best chance of surviving if it is short. For example, in 338 B.C. Philip of Macedon, at the height of his power, swept into southern Greece. The Greek cities submitted at once, all except Sparta. Sparta had been the foremost military power in Greece until 371 B.C., but since then had declined into near helplessness. Philip, annoyed by the defiance of this impotent city, sent this message: "If I attack Sparta, I will level it to the ground." Back at once came the reply of a single word: "If!" Philip, admiring the spirit of the city, left without brandishing a weapon.

Another way of preserving a quotation is to present it as poetry. It is easier to remember a passage that contains rhythm and rhyme than one that does not. Consider the man whom many people deem to be the greatest scientist in history. His greatness might be spoken of in paragraphs and pages and yet not be captured as well as Alexander Pope managed to do in two lines of verse intended as an epitaph for the great man:

> *Nature and Nature's laws lay hid in night:*
> *God said, "Let Newton be" and all was light.*

William Wordsworth commented on a statue of Newton:

> *The marble index of a mind forever*
> *Voyaging through strange seas of thought, alone.*

That powerful last word, "alone," emphasizes the uniqueness of the man and makes the remark an unforgettable quotation.

But what did Newton himself say that would be suitable as a quotation? Naturally, the greatest scientific mind in all history cannot be expected to boast of it:

> *To myself I seem to have been only like a boy*
> *playing on the seashore, and diverting*
> *myself in now and then finding a smoother*
> *pebble or a prettier shell than ordinary,*
> *whilst the great ocean of truth lay all*
> *undiscovered before me.*

That is not verse, and it is too long for the average person to remember, but when one reads it, what a picture it presents, and how it forces the myriads of lesser minds in the world to long to wrest one drop of truth from that great ocean.

If we speak of scientific quotations, particularly, one has the chance of surviving if it expresses, most pithily, some great truth. We know, for instance, that everything moves. A rock may seem motionless, but every rock and every mountain is moving with the Earth around the planetary axis, and also about the Sun, and also about the galactic center. Within the rock there are atoms and molecules that restlessly and eternally vibrate, and very slowly shift position. On a more visible scale, the ocean waves, the river flows, living things age, die, and decay. How best to say all this? The ancient Greek philosopher Heracleitus put it this way: "It is not possible to step twice into the same river." Think about it. You step into new water each time.

Scientists have always tried to simplify matters. They have tried to find very broad generalizations that explain as much as possible in as compact a way as possible. They try to cut away the superficial and trivial and get down to basics. As the ancient Greek philosopher Democritus concluded, "There is nothing but atoms and the void." By atoms, he meant fundamental particles, but there are also fields of force (such as gravitation) that fill all space, so that a true void does not exist without them; you can, therefore, still argue that Democritus was correct. And the conciseness of the saying gives you an image of the Universe that is pregnant with many philosophical implications.

Ultimately, what Democritus was saying was that if we could understand the Universe deeply enough, we would find it was simple. This may be false, but in the 2,500 years since Democritus, scientific findings have strongly supported the notion of simplicity. The Universe is made up of two dozen fundamental particles and four fields, and scientists are busily engaged in trying to reduce even these to varieties of a single particle and a single field—and may succeed.

Whereas Heracleitus and Democritus are believed to be fundamentally correct in their quotations, it is possible for a great mind to be wrong (or *appear* to be wrong) and yet express that view in so forceful and picturesque a manner that it lives as a quotation and, as a quotation, nearly conquers the truth. The twentieth century has produced two great theories that rule all of physics now. One is the theory of relativity, developed by Albert Einstein in 1905 and brilliantly extended in 1916. The other is quantum theory, first advanced by Max Planck in 1900, but first demonstrated to be useful by, again, Einstein, in 1905.

Quantum theory was reduced to a strict mathematical formalism in the 1920's by men such as Erwin Schrodinger and Werner Heisenberg. This "quantum mechanics," however, introduced the principle of uncertainty into physics. One could not, even in principle, determine all the properties of a particle to any desired accuracy. The size of Planck's constant produces an ultimate "fuzziness" to the Universe that seems to vitiate causality. An electron, under the same impulse, might go here or it might go there; there was no predicting.

Einstein would not accept this aspect of quantum theory. To the end of his life, he felt that quantum theory was incomplete and that there was a deeper, more complete theory that would restore causality and make it quite plain where the electron would go, leaving no room for uncertainty. In this, Einstein seems to be wrong. Since quantum mechanics has been formulated, it has successfully met all challenges and has successfully predicted the probabilities of events with great accuracy.

Nevertheless, this is what Einstein said, with respect to the notion that events proceed according to probabilities only: "I am convinced that God does not play dice." The image is so forceful that even those who accept quantum mechanics are forced to do so uneasily.

All science is based on two assumptions. One is that the Universe can be explained by evidence obtained from the Universe alone; that no supernatural agency need be called upon. The other is that the human mind is, in the long run, capable of understanding the Universe.

These are only assumptions, and no one can prove they are true. However, working on these assumptions, science has accomplished great things so that scientists feel increasingly confident, as the decades pass, that these assumptions *are* true. And just as Einstein makes us uncomfortable about quantum mechanics with his quotation about God playing dice, he comforts us about the assumption of the understandability of the Universe with another well-known quotation about God: "The Lord God is subtle, but malicious he is not." In other words, God has not made it easy to understand the Universe, but neither has he deliberately planted false or misleading evidence merely to confuse us.

You see, then, the potential importance of quotations in the field of science. They are metaphors of a reality that might otherwise be difficult to grasp. They are shortcuts to an inner wisdom. They inspire students and scientists alike. And they enlarge the worldview and help give rise to further questions.

Does the quotation express a lasting truth, or do time and change make it obsolete? If it is true, what is the basic truth it encapsulates, and of what use is it to science as a whole? You may surprise yourself with the byways of thought into which the quote might lead you, and the subtle pleasures you will derive from it.

Then, too, we must be cautious with quotations. Quotations are tricky. Occasionally they are apocryphal. We sometimes do not have them directly from the writings of the person supposed to have said them. They may come instead from the writings of someone else who lived decades or even centuries later. They may be recalled incorrectly, or they may belong to that class of quotation which the person might not have said but which someone else felt he *should* have said.

For example, it is widely repeated that Galileo, after agreeing at the order of the Inquisition no longer to teach or believe that the Earth revolves about the Sun, or moves in any way, stamped his foot and muttered under his breath: "*Eppur si muove*" (Just the same, it moves). It is very doubtful that Galileo really said that, but the statement fits his character and is a beautiful commentary on the uselessness of political decisions on what ought to be believed. In other words, if Galileo did not say it, he *should* have.

Then, too, a quotation may be, originally, in a foreign language and in need of translation before it can be included here—and the translation may be inaccurate or distorted.

These are sins of commission, but sins of omission may be just as bad, if not necessarily as noticeable. We may have left out some excellent quotation through ignorance or carelessness.

If you have reason to believe that a quotation was never said or written, or that it is quoted incorrectly, or translated inadequately, please let us know. If you know of any quotations that you think are delightful, try them out on us. We hope that there will be future editions of this book, and we have no objections to having those editions improved by our audience, as well as by our own further efforts.

As it stands now, the book is presented to its audience as a way of epitomizing views on science, presenting wisdom in brief, inspiring readers with curiosity about the wonderful Universe, and granting them, perhaps, an added insight into the delights of science.

NEW YORK CITY
1987

Introduction

BY JASON A. SHULMAN

Real knowledge goes into natural man in tidbits. A scrap here, a scrap there; always pertinent, linked to safety or nutrition or pleasure.

EZRA POUND

I HAVE always loved the way the world is made. From the tiny swirls of pinecones to the swirling clouds of Jupiter, nature has a rightness, a sense, that always simultaneously enlightens and baffles my mind.

I grew up in love with science, asking the same questions all children ask as they try to codify the world to find out what makes it work. "Who is the smartest person in the world?" and "Where is the tallest mountain in the world?" turned into questions like, "How big is the Universe?" and "What is it that *makes* us alive?"

Although we all grow up, and these questions give way to more sophisticated ones, I believe that all science is based on the spirit of these original queries, and that they remain with us always and live, the foundation of all of our quests, like multicolored pebbles on the bed of a fast-flowing stream. We spend our eighty years or so feeling for the pebbles with our toes, trying to gather enough information to make a coherent picture of the world.

When we are young, we think that science has to do with facts, with finding answers and solutions, and that it proceeds like an arrow from the primitive to the sophisticated, from mystery to light. But as we get older, we find that, while science does have to do with facts and laws, it has equally to do with wisdom, which is something else entirely. The wisdom of science knows no boundaries and does not proceed in an orderly manner from past to present, always building on existing knowledge. Instead, it appears throughout time, regardless of mere technical sophistication. It appears in Claude Bernard's words from the mid-1800's:

> The true worth of an experimenter consists in his pursuing not only what he seeks in his experiment, but also what he did not seek.

. . . and Einstein's, from almost one hundred years later:

> The mere formulation of a problem is often far more essential than its solution, which may be merely a matter of mathematical or experimental skill. To raise new questions, new possibilities, to regard old problems from a new angle requires creative imagination and marks real advances in science.

We learn that while there are scientists who devote their lives to the study of the subject, the concerns of science—far from being limited in interest only to scientists—stretch like a web throughout our completely interrelated Universe, daily affecting all of our lives in profound ways. The old notion of science as a description of how we live is really linked with the profound question of why we live as well.

For me, the facts of science are only way stations on the path to something else. Science itself is just our way of finding out what nature already knows: we create nothing ourselves, we simply discover deeper applications of natural laws and make use of them in the presence or absence of wisdom.

Many of the major discoveries of scientific laws were made by scientists who took the natural world as their starting place. This was certainly true of Newton, whose profound investigations into light and gravity were based first on observation. Even Einstein's Special Theory of Relativity was prompted by his recognition of natural, everyday phenomena.

In all, nature is our teacher, and science does not move a step without her. When we commune with that level of the world, we become the "natural" men and women the poets talk about; we become the best of scientists.

In recognition of that reality, this book is not so much a book of science fact as a book of science wisdom, the place where humankind's fact gathering bears fruit and resembles nature most of all.

Quotations are the perfect way to encapsulate science's wisdom, and in this book you will find over two thousand of them—quotations that will move you, scientist or not, to see that the world is indeed interrelated, and that the job of science is to find our place in it.

These are bits of our best. May they be food to the child in you, the one who has never given up asking, "What is it that makes us live?" and "How big is infinity?"

BROOKLYN
1987

About This Book

THIS BOOK is a short history of ideas. Hence the quotations in this book are arranged chronologically within each category, either by the birth date of the speaker or by the date of the quotation itself. Because the older material is difficult to date precisely, we have dated those quotations by the birth and death dates of the speaker, since those are readily available. The dates of the newer quotations are much easier to gauge accurately, and since five or ten years can represent a great change in attitude toward the particular point of view expressed, the arrangement is by the date of the quote itself.

Biblical material is divided into Old Testament and New Testament entries, with the Old Testament entries dated 725 B.C. and New Testament entries dated 325 A.D. These dates are approximate, but for our purposes, allow us to put Genesis before Aristotle and the epistles of Paul after Horace.

Each entry provides a brief description of the speaker. These descriptions are of necessity short and are intended to help provide context for the quote.

1.
Aeronautics

Surely no child, and few adults, have ever watched a bird in flight without envy.

ISAAC ASIMOV

1.1 Oh that I had wings like a dove! for then would I fly away, and be at rest.
***The Bible* (circa 725 B.C.)**

1.2 A bird is an instrument working according to mathematical law, which instrument it is within the capacity of man to reproduce with all its movements, but not with a corresponding degree of strength, though it may therefore say that such an instrument constructed by man is lacking in nothing except the life of the bird, and this life must needs be supplied from that of man.
Leonardo da Vinci, Italian Architect/Artist/Inventor (1452–1519)

1.3 The [mechanical] bird I have described ought to be able by the help of the wind to rise to a great height, and this will prove to be its safety; since even if . . . revolutions [of the winds] were to befall it, it would still have time to regain a condition of equilibrium; provided that its various parts have a great power of resistance, so that they can safely withstand the fury and violence of the descent, by the aid of the defenses which I have mentioned; and its joints should be made of strong tanned hide, and sewn with cords of strong raw silk. And let no one encumber himself with iron bands, for these are very soon broken at the joints or else they become worn out, and consequently it is well not to encumber oneself with them.
Leonardo da Vinci, Italian Architect/Artist/Inventor (1452–1519)

1.4 If a man has a tent made of linen of which the apertures have all been stopped up, and it be twelve braccia across (over twenty-five feet) and twelve in

depth, he will be able to throw himself down from any height without sustaining injury.
[On the invention of the parachute.]
Leonardo da Vinci, Italian Architect/Artist/Inventor (1452–1519)

1.5 What can you conceive more silly and extravagant than to suppose a man racking his brains, and studying night and day how to fly?
William Law, English Clergyman/Writer (1686–1761)

1.6 Soon shall thy arm
Unconquer'd steam! Afar
Drag the slow barge or drive the rapid car,
Or on wide-waving wings expanded bear
The flying chariot through the fields of air.
Erasmus Darwin, English Physician/Poet (1731–1802)

1.7 We will build a machine that will fly.
[He invented the first air balloon, 1784.]
Joseph Michael Montgolfier, French Inventor/Aeronaut (1740–1810)

1.8 The late Mr. Sadler, the celebrated aëronaut, ascended on one occasion in a balloon from Dublin, and was wafted across the Irish Channel, when, on his approach to the Welsh coast, the balloon descended nearly to the surface of the sea. By this time the sun was set, and the shades of evening began to close in. He threw out nearly all his ballast, and suddenly sprang upwards to a great height, and by so doing brought his horizon to *dip* below the sun, producing the whole phenomenon of a western sunrise.
Sir John Herschel, English Astronomer (1792–1871)

1.9 All at once, but with such rapidity and to such a prodigious elevation that we had difficulty in hearing each other—even when shouting at the top of our voices. I was ill and vomited; Grassetti was bleeding at the nose. We were both breathing short and hard, and felt oppression on the chest. Because the balloon rose so suddenly out of the water and bore us with such swiftness to those high regions, the cold seized us suddenly, and we found ourselves covered with a layer of ice.
Count Francesco Zambeccari, Italian Aeronaut (1804)

1.10 "What is the news, good neighbor, I pray?"
"They say a balloon has gone up to the moon
And won't be back till a week from today."
Nursery Rhyme (circa 1805)

1.11 The feeling of absolute solitude is rarely experienced upon earth, but in these regions, separated from all human associations, the soul might almost fancy it had passed the confines of the grave. Nature was noiseless, even the wind was silent, therefore, receiving no opposition, we gently floated along, and the lonely stillness was interrupted only by the progress of the car and its

colossal ball which, self-propelled, seemed like the rockbird fluttering in the blue ether.
[On flight in a hot air balloon, 1817.]

Prince Hermann Ludwig Heinrich von Pückler-Muskau, German Prince (1785–1871)

1.12 Another contest had been announced. Enthusiastic competitors arrived with their streamlined "birds" [gliders] of all colours and types. The slightly thundery atmosphere gave promise of record-breaking attempts, and in fact heights of 26,000 feet were reached in several cases. Then five daring contestants flew into a thundercloud. . . . We do not know all that happened between earth and sky during these frightful minutes, as only one man, severely injured, escaped with his life. We can only imagine the ordeal of the four. At a height of thirty-five, forty-five, or even fifty thousand feet they must have been enclosed in a casing of frozen water, tossed about like living icicles, stabbed at by lightning, until the cloud released their four lifeless bodies.

Leo Loebsack, German Science Writer (circa 1817)

1.13 The birds can fly,
An' why can't I?

John Townsend Trowbridge, American Novelist/Poet (1827–1916)

1.14 Men will never fly, because flying is reserved for angels.

Bishop Milton Wright, American Episcopalian Bishop/Father of the Wright brothers (1828–1917)

1.15 And this is aviation; I give it to the world.

Louis Mouillard, French Inventor/Aeronaut (1834–1897)

1.16 We were on the point of abandoning our work when the book of Mouillard fell into our hands, and we continued with the results you know.

Wilbur Wright, American Inventor/Aviator (1867-1912)

1.17 It is possible to fly without motors, but not without knowledge and skill.

Wilbur Wright, American Inventor/Aviator (1867–1912)

1.18 Aeronautics must make many a long stride before they do much practical work, either in commerce or war.

***Echo*, English periodical (October 19, 1870)**

1.19 Successful—four flights on Thursday morning—took off with motors from level ground—average speed thirty miles an hour—longest flight 59 seconds—inform press—home for Christmas—Orville.
[His telegram to his father, Bishop Wright, about the first flight in an airplane, Kitty Hawk, N.C., December 17, 1903.]

Orville Wright, American Inventor/Aviator (1871–1948)

1.20 The wildest stretch of imagination of that time would not have permitted us to believe that within a space of fifteen years actually thousands of these machines would be in the air engaged in deadly combat.

Orville Wright, American Inventor/Aviator (1871–1948)

1.21 In ancient days two aviators procured to themselves wings. Daedalus flew safely through the middle air and was duly honored on his landing. Icarus soared upwards to the sun till the wax melted which bound his wings and his flight ended in fiasco. In weighing their achievements, there is something to be said for Icarus. The classical authorities tell us that he was only "doing a stunt," but I prefer to think of him as the man who brought to light a serious constructional defect in the flying machines of his day.

Sir Arthur Stanley Eddington, English Astronomer/Mathematician (1882–1944)

1.22 I would much prefer to have Goddard interested in real scientific development than to have him primarily interested in more spectacular achievements [Goddard's rocket research] of less real value.

Charles Augustus Lindbergh, American Aviator (1902–1974)

1.23 We hope that Professor Langley will not put his substantial greatness as a scientist in further peril by continuing to waste his time, and the money involved, in further airship experiments. Life is too short, and he is capable of services to humanity incomparably greater than can be expected to result from trying to fly. . . . For students and investigators of the Langley type there are more useful employments.

***The New York Times* (1903)**

1.24 Bombardment from the air is legitimate only when directed at a military objective, the destruction or injury of which would constitute a distinct military disadvantage to the belligerent.

The Second Hague Peace Conference (1907)

1.25 There are . . . recurrent and apparently reliable reports that Pan American in conjunction with British Imperial Airways is soon to put into operation transatlantic and New York–Bermuda airlines.

***Scientific American* (1936)**

1.26 It's burst into flames! Oh my . . . it's burning, bursting into flames! . . . oh the humanity and all the passengers!
[Live broadcast of the Hindenberg disaster.]

Herbert Morrison, American Radio Journalist (1937)

1.27 In the space age, man will be able to go around the world in two hours—one hour for flying and the other to get to the airport.

Neil McElroy, American Business Executive (1958)

1.28 The common denominator in achieving successful translation of aerospace technology into the commercial sector is the need for a government program with a definite objective and an interested sponsor or customer for their services or products to be produced.

Robert Anderson, American Industrialist (1971)

1.29 Asking an aerospace worker if he's ever been laid off before is like asking a mother if she's ever had a baby.

Richard Kapusta, American Aerospace Educator (1971)

1.30 We hear the airlines say "We ought to have the right of way." That's like Greyhound demanding that all cars get out of its way on the highway.

John Baker, American Aviator (1986)

2.
Agriculture

When the practice of farming spread over the earth,

mankind experienced its first population explosion.

ISAAC ASIMOV

2.1 And God said, "Let the earth put forth vegetation, plants yielding seed, and fruit trees bearing fruit in which is their seed, each according to its kind." And God saw that it was good. And there was evening and there was morning, a third day.

***The Bible* (circa 725 B.C.)**

2.2 Agriculture, for an honorable and high-minded man, is the best of all occupations or arts by which men procure the means of living.

Xenophon, Greek Military Leader/Historian (431 B.C.?–352 B.C.?)

2.3 A farmer is always going to be rich next year.

Philemon, Roman Comic Poet (365 B.C.?–265 B.C.?)

2.4 A field becomes exhausted by constant tilling.

Ovid (Publius Ovidius Nasso), Roman Poet (43 B.C.–17 A.D.?)

2.5 The earth conceives by the sun, and through him becomes pregnant with annual fruits.

Nicholas Copernicus, Polish Astronomer (1473–1543)

2.6 The frost is God's plough which he drives through every inch of ground in the world, opening each clod, and pulverizing the whole.

Thomas Fuller, English Historian/Theologian (1608–1661)

2.7 The first three men in the world were a gardener, a ploughman, and a grazier; and if any object that the second of these was a murderer, I desire him to consider that as soon as he was so, he quitted our profession, and turned builder.

Abraham Cowley, English Poet (1618–1667)

2.8 He who appropriates land to himself by his labor, does not lessen but increases the common stock of mankind. For the provisions serving to the support of human life, produced by one acre of inclosed and cultivated land, are . . . ten times more than those which are yielded by an acre of land, of an equal richness lying waste in common. And therefore he that incloses land and has a greater plenty of the conveniences of life from ten acres than he could have from a hundred left to nature, may truly be said to give ninety acres to mankind.

John Locke, English Philosopher (1632–1704)

2.9 In another Apartment I was highly pleased with a Projector, who had found a Device of plowing the Ground with Hogs, to save the Charges of Plows, Cattle, and Labour. The Method is this: In an Acre of Ground you bury at six Inches Distance, and eight deep, a Quantity of Acorns, Dates, Chestnuts, and other Masts or Vegetables whereof these Animals are fondest; then you drive six Hundred or more of them into the Field, where in a few Days they will root up the whole Ground in search of their Food, and make it fit for sowing, at the same time manuring it with their Dung. It is true, upon Experiment they found the Charge and Trouble very great, and they had little or no Crop. However, it is not doubted that this Invention may be capable of great Improvement.

Jonathan Swift, Irish Satirist/Clergyman (1667–1745)

2.10 There seem to be but three ways for a nation to acquire wealth: the first is by war, as the Romans did, in plundering their conquered neighbors—this is robbery; the second by commerce, which is generally cheating; the third by agriculture, the only honest way, wherein man receives a real increase of the seed thrown into the ground, in a kind of continual miracle, wrought by the hand of God in his favor, as a reward for his innocent life and his virtuous industry.

Benjamin Franklin, American Inventor/Statesman (1706–1790)

2.11 Agriculture not only gives riches to a nation, but the only riches she can call her own.

Samuel Johnson, English Lexicographer/Poet/Critic (1709–1784)

2.12 Agriculture is the foundation of manufactures, since the productions of nature are the materials of art.

Edward Gibbon, English Historian (1737–1794)

2.13 Let the farmer forevermore be honored in his calling, for they who labor in the earth are the chosen people of God.

Thomas Jefferson, American President/Author (1743–1826)

2.14 Farming . . . is commensurate with the postures of the female mind; nor is the practice of inspecting agricultural processes, incompatible with the delicacy of their frames, if their constitution is good.

Priscilla Wakefield, English Author/Philanthropist (1751–1832)

2.15 When tillage begins, other arts follow. The farmers, therefore, are the founders of human civilization.

Daniel Webster, American Statesman/Orator (1782–1852)

2.16 In a moral point of view, the life of the agriculturist is the most pure and holy of any class of men; pure, because it is the most healthful, and vice can hardly find time to contaminate it; and holy, because it brings the Deity perpetually before his view, giving him thereby the most exalted notions of supreme power, and the most endearing view of the divine benignity.

Lord John Russell, English Statesman/Author (1792–1878)

2.17 Earth is so kind, that just tickle her with a hoe and she laughs with a harvest.

Douglas William Jerrold, English Humorist (1803–1857)

2.18 In order to civilize a people, it is necessary first to fix it, and this cannot be done without inducing it to cultivate the soil.

Alexis de Tocqueville, French Political Scientist/Historian (1805–1859)

2.19 He that would look with contempt on the pursuits of the farmer, is not worthy of the name of a man.

Henry Ward Beecher, American Preacher/Author (1813–1887)

2.20 Blessed be agriculture! If one does not have too much of it.

Charles Dudley Warner, American Novelist/Essayist (1829–1900)

2.21 The farmer works the soil,
The agriculturist works the farmer.

Eugene Fitch Ware ("Ironquill"), American Lawyer/Verse Writer (1841–1911)

2.22 The farther we get away from the land, the greater our insecurity.

Henry Ford, American Industrialist/Auto Maker (1863–1947)

2.23 The oranges, it is true, are not all exactly of the same size, but careful machinery sorts them so that automatically all those in one box are exactly similar. They travel along with suitable things being done to them by suitable machines at suitable points until they enter a suitable refrigerator car in which they travel to a suitable market. The machine stamps the word "Sunkist" upon them, but otherwise there is nothing to suggest that nature has any part in their production.

Bertrand Russell, English Philosopher/Mathematician (1872–1970)

2.24 In the Soviet Union farmers look in the barn for "their" horses even after they have given them to the collective.

Nikita Sergeevich Khrushchev, Soviet Statesman/Prime Minister (1894–1971)

2.25 The chief problem of the lower-income farmers is poverty.

Nelson Aldrich Rockefeller, American Politician/Vice President (1908–1979)

2.26 I want action, not talk, a decent price for what I raise, not a handout.

Edwin Dent, American Farmer (1971)

2.27 By 1980, two-thirds of our present farmers, who have been raised in the tradition of owning their equipment, will be retired. I think we have to expect that the young, aggressive, well-educated operators, who are already beginning to assume the management of agriculture, will weigh very carefully the advantages of leasing or renting their equipment.

Brooks McCormick, American Agricultural Industrialist (1971)

2.28 Peace means creative giving, making a contribution to the betterment of our own people and to the betterment of people throughout the world. And who better understands that than the men and women of American agriculture?

Richard Milhous Nixon, American President (1971)

2.29 Food is a weapon . . . one of the principal tools in our negotiating kit.

Earl Butz, American Secretary of Agriculture (1975)

2.30 A third of all cropland is suffering soil losses too great to be restrained without a gradual, but ultimately disastrous, decline in productivity.

Council for Agricultural Science (1976)

2.31 In a hundred and fifty years the United States has lost one third of its topsoil. And I think about two hundred fifty million acres are turning into desert because of overgrazing and other mismanagement.

George Schaller, American Zoologist/Ecologist/Educator (1984)

2.32 We have a harmless organism. We put a gene into it that produces a toxin that has been used for 20 years without any problems at all. We completed test after test to prove it was safe. What do we need to do to convince people that planting one-tenth of an acre of treated corn is not going to ruin the world? [Regarding a controversy over bioengineering research for breeding pest-resistant plants.]

Robert J. Kaufman, American Biotechnology Researcher (1986)

3.

Anatomy

The human mechanism is marvelous. But why not—it is the result of three-and-a-half billion years of tinkering.

ISAAC ASIMOV

3.1 The heart is an exceedingly strong muscle. . . . It contains two separate cavities.

Hippocrates, Greek Physician (460 B.C.?–377 B.C.?)

3.2 The heart in all animals has cavities inside it. . . . The largest of all the three chambers is on the right and highest up; the least is on the left; and the medium one lies in between the other two.

Aristotle, Greek Philosopher (384 B.C.–322 B.C.)

3.3 It is not impossible to obtain a view of human bones. . . . Once I examined the skeleton of a robber, lying on a mountain-side a short distance from the road. This man had been killed by some traveler whom he had attacked, but who had been too quick for him. None of the inhabitants of the district would bury him; but in their detestation of him they were delighted when his body was eaten by birds of prey; the latter, in fact, devoured the flesh in two days and left the skeleton ready, as it were, for anyone who cared to enjoy an anatomical demonstration.

Claudius Galen, Greek Physician/Scholar (130–200)

3.4 The heart is the beginning of life; the sun of the microcosm, even as the sun in his turn might well be designated the heart of the world; for it is the heart by whose virtue and pulse the blood is moved, perfected, made apt to nourish, and is preserved from corruption and coagulation; it is the household divinity which, discharging its function, nourishes, cherishes, quickens the whole body, and is indeed the foundation of life, the source of all action. . . . The heart, like the prince in a kingdom, in whose hands lie the chief highest authority, rules over all; it is the original and foundation from which all power is derived, on which all power depends in the animal body.

William Harvey, English Anatomist/Physician (1578–1657)

3.5 The soul seems to be a very tenuous substance (and) seems to be made of a most subtle texture, extremely mobile or active corpuscles, not unlike those of flame or heat; indeed, whether they are spherical, as the authors of atoms propound, or pyramidical as Plato thought, or some other form, they seem from their own motion and penetration through bodies to create the heat which is in the animal.

Pierre Gassendi, French Physicist/Philosopher/Priest (1592–1655)

3.6 The anatomy of a little child, representing all parts thereof, is accounted a greater rarity than the skeleton of a man in full stature.

Thomas Fuller, English Historian/Theologian (1608–1661)

3.7 In vain we shall penetrate more and more deeply the secrets of the structure of the human body, we shall not dupe nature; we shall die as usual.

Bernard Le Bovier Sieur de Fontenelle, French Philosopher (1657–1757)

3.8 Like following life thro' creatures you dissect,
You lose it in the moment you detect.

Alexander Pope, English Poet/Satirist (1688–1744)

3.9 No anatomist ever discovered a system of organization calculated to produce pain and disease; or, in explaining the parts of the human body, ever said, this is to irritate; this is to inflame; this duct is to convey the gravel to the kidneys; this gland to secrete the humour which forms the gout: if by chance he come at a part of which he knows not the use, the most he can say is, that it is useless; no one ever suspects that it is put there to incommode, to annoy, or to torment.

William Paley, English Educator/Theologian (1743–1805)

3.10 No knowledge can be more satisfactory to a man than that of his own frame, its parts, their functions and actions.

Thomas Jefferson, American President/Author (1743–1826)

3.11 A man is as old as his arteries.

Pierre Jean George Cabanis, French Physician/Author (1757–1808)

3.12 The curious adaptation of a member to different offices and to different conditions of the animal has led to a very extraordinary opinion in the present day—that all animals consist of the same elements.

Sir Charles Bell, Scottish Anatomist/Surgeon (1774–1842)

3.13 The Chinese, who aspire to be thought an enlightened nation, to this day are ignorant of the circulation of the blood; and even in England the man who made that noble discovery lost all his practice in the consequence of his ingenuity; and Hume informs us that no physician in the United Kingdom who had attained the age of forty ever submitted to become a convert to Harvey's theory, but went on preferring *numpsimus* to *sumpsimus* to the day of his death.

Charles Caleb Colton, English Clergyman/Writer (1780?–1832)

3.14 Anatomists see no beautiful woman in all their lives, but only a ghastly sack of bones with Latin names to them, and a network of nerves and muscles and tissues inflamed by disease.

Mark Twain (Samuel Langhorne Clemens), American Journalist/ Novelist (1835–1910)

3.15 You remember your promise? You will do my post-mortem? And look at the intestines carefully, for I think there is something there now.
[His last words.]

Ilya Ilich Metchnikov, Russian Zoologist (1845–1916)

3.16 We are accustomed to say that every human being displays both male and female instinctual impulses, needs, and attributes, but the characteristics of what is male and female can only be demonstrated in anatomy, and not in psychology.

Sigmund Freud, Austrian Psychiatrist/Psychoanalyst (1856–1939)

3.17 Neither the absolute nor the relative size of the brain can be used to measure the degree of mental ability in animal or in man. So far as man is concerned, the weights of the brains or the volumes of the cranial cavities of a hundred celebrities of all branches of knowledge all over the world have been listed. At the bottom of those lists are Gall, the famous phrenologist, Anatole France, the French novelist, and Gambetta, the French statesman, each with about 1,100 cc brain mass. The lists are topped by Dean Jonathan Swift, the English writer, Lord Byron, the English poet, and Turgenev, the Russian novelist, all with about 2,000 cc . . . Now our mental test! Had Turgenev really twice the mental ability of Anatole France?

Franz Weidenreich, German/American Anatomist/Physical Anthropologist (1873–1948)

4.
Animal Behavior

To insult someone we call him "bestial."

For deliberate cruelty and malice, "human"

might be the greater insult.

ISAAC ASIMOV

4.1 Now the serpent was more subtle than any beast of the field.
The Bible (circa 725 B.C.)

4.2 When a building is about to fall down, all the mice desert it.
Pliny the Younger (Gaius Plinius Caecilius Secundus), Roman Naturalist/Scholar (62–113)

4.3 The bee is more honored than other animals, not because she labors, but because she labors for others.
Saint John Chrysostom, Greek Church Father (345–407)

4.4 An ape is zany to a man, doing over those tricks (especially if they be knavish) which he sees done before him.
Thomas Dekker, English Dramatist/Pamphleteer (1572–1632)

4.5 Of beasts, it is confess'd, the ape
Comes nearest us in human shape;
Like man he imitates each fashion,
And malice is his ruling passion.
Jonathan Swift, Irish Satirist/Clergyman (1667–1745)

4.6 Animal enjoyments are infinitely *diversified*.
William Paley, English Educator/Theologian (1743–1805)

4.7 Besides love and sympathy, animals exhibit other qualities connected with the social instincts which in us would be called moral.

Charles Robert Darwin, English Naturalist/Evolutionist (1809–1882)

4.8 Animals are such agreeable friends; they ask no questions, pass no criticisms.

George Eliot (Mary Ann Evans), English Novelist (1819–1880)

4.9 No human being was ever so free as a fish.

John Ruskin, English Art Critic/Author (1819–1900)

4.10 Cats and monkeys, monkeys and cats—all human life is there.

Henry James, Jr., American Novelist/Critic (1843–1916)

4.11 A vast number, perhaps the numerical majority, of animal forms cannot be shown unequivocally to possess mind.

Sir Charles Scott Sherrington, English Physiologist (1857–1952)

4.12 For my part, I am not so sure at bottom that man is, as he says, the king of nature; he is far more its devastating tyrant. I believe he has many things to learn from animal societies, older than his own and of infinite variety.

Romain Rolland, French Novelist (1866–1944)

4.13 It seemed that animals always behave in a manner showing the rightness of the philosophy entertained by the man who observes them. . . . Throughout the reign of Queen Victoria all apes were virtuous monogamists, but during the dissolute twenties their morals underwent a disastrous deterioration.

Bertrand Russell, English Philosopher/Mathematician (1872–1970)

4.14 Sparrows are sociable, like a crowd of children begging from a tourist.

Robert Staughton Lynd, American Sociologist (1892–1970)

4.15 Meat-eating has not, to my knowledge, been recorded from other parts of the chimpanzee's range in Africa, although if it is assumed that human infants are in fact taken for food, the report that five babies were carried off in West Africa suggests that carnivorous behavior may be widespread.

Jane Goodall, American Primatologist (1934–)

4.16 The Gombe Stream chimpanzees . . . in their ability to modify a twig or stick to make it suitable for a definite purpose, provide the first examples of free-ranging nonhuman primates actually *making* very crude tools.

Jane Goodall, American Primatologist (1934–)

4.17 It doesn't take a scientist to realize that a chimpanzee or a dog is an intelligent animal. Instead, it takes a bigoted human to suggest that it's not.

Richard Leakey, English Paleontologist (1944–)

4.18 Anybody who says he's been et by a wolf is a liar.

John B. Theberge, American Ecologist/Zoologist (1971)

4.19 Ants are so much like human beings as to be an embarrassment. They farm fungi, raise aphids as livestock, launch armies into wars, use chemical sprays to alarm and confuse enemies, capture slaves. . . . They exchange information ceaselessly. They do everything but watch television.

Lewis Thomas, American Medical Educator/Writer (1974)

4.20 I did some testing a long time ago and showed that porpoises distinguish between Strauss's "Vienna Woods" waltz and all other Strauss waltzes. They had been fed to it. When I played other waltzes, they ignored me, but when I played "Vienna Woods," they started jumping around the platform.

Karl Pribram, Austrian/American Neurophysiologist (1984)

4.21 I suggest you don a dolphin suit and join them. It's an incredible experience to hang out with them.

John Lilly, American Neurophysiologist/Spiritualist (1984)

4.22 Elephants love bathing, and there are numerous reports of them island hopping in the open sea off India and Sri Lanka (their trunks make excellent snorkels).

Paul Y. Sondaar, Dutch Paleontologist/Educator (1986)

4.23 The key [to animal communication] is what the sounds mean. If you get two roars—one after sex and one during eating—is there anything in the acoustics of those roars that means something specific?

Harry Hollien, American Linguist (1986)

4.24 Llamas mate sitting down. That is probably reason enough to study them.

Gamini Seneviratne, Science Journalist (1986)

5.
Anthropology

Human societies are everywhere complex,

for living at peace with ourselves requires

a vast multiplicity of rules.

ISAAC ASIMOV

5.1 Aristippus, a companion of Socrates, shipwrecked and cast up on the shore of Rhodes, noted geometrical figures drawn on the sand. "Let us be in good hope, for indeed I see the traces of man."
Marcus Vitruvius Pollio (Vitruvius), Roman Architect/Military Engineer (circa 100 B.C.)

5.2 The man is not of the woman; but the woman of the man. Neither was the man created for the woman; but the woman for the man.
***The Bible* (circa 325 A.D.)**

5.3 I have spent much time in the study of the abstract sciences; but the paucity of persons with whom you can communicate on such subjects disgusted me with them. When I began to study man, I saw that these abstract sciences are not suited to him, and that in diving into them, I wandered farther from my real object than those who knew them not, and I forgave them for not having attended to these things. I expected then, however, that I should find some companions in the study of man, since it was so specifically a duty. I was in error. There are fewer students of man than of geometry.
Blaise Pascal, French Mathematician/Philosopher/Author (1623–1662)

5.4 Man is an animal and should live in accordance with his species.
Carl Linnaeus, Swedish Botanist/Naturalist (1707–1778)

5.5 Wild man: four-footed, mute and hairy.
American man: erect, choleric, obstinate, gay, free. Lives by custom.
European man: gentle, clever, inventive. Governed by rites.
Asiatic man: melancholy, rigid, severe, discriminating, greedy. Governed by opinions.
African man: crafty, indolent, and negligent. Governed by price.
[His classification of human species.]

Carl Linnaeus, Swedish Botanist/Naturalist (1707–1778)

5.6 Three ways have been taken to account for it [racial differences]: either that they are the posterity of Ham, who was cursed; or that God at first created two kinds of men, one black and another white; or that by the heat of the sun the skin is scorched, and so gets the sooty hue. This matter has been much canvassed among naturalists, but has never been brought to any certain issue.

Samuel Johnson, English Lexicographer/Poet/Critic (1709–1784)

5.7 People will not look forward to posterity, who never look backward to their ancestors.

Edmund Burke, British Statesman/Author (1729–1797)

5.8 There are no inferior races; all are destined to attain freedom.

Friedrich Heinrich Alexander von Humboldt, German Naturalist/Statesman (1769–1859)

5.9 Man is a tool-using animal.

Thomas Carlyle, Scottish Historian/Philosopher (1795–1881)

5.10 Man is the highest product of his own history. The discoverer finds nothing so grand or tall as himself, nothing so valuable to him. The greatest star is at the small end of the telescope, the star that is looking, not looked after nor looked at.

Theodore Parker, American Clergyman/Author (1810–1860)

5.11 In no sense can the Neanderthal bones be regarded as the remains of a human being intermediate between men and apes.

Thomas Henry Huxley, English Biologist/Evolutionist (1825–1895)

5.12 Men make the gods; women worship them.

Sir James George Frazer, Scottish Anthropologist (1854–1941)

5.13 There is neither East nor West, Border nor Breed nor Birth, when two strong men stand face to face, though they come from the end of the earth.

Rudyard Kipling, English Author/Poet (1865–1936)

5.14 But although these numerous traces of prehistoric man found in America might lead us to suppose that there was the birthplace of the human race, we are unable to support that theory.

Samuel Waddington, American Writer (1900)

5.15 Anthropologists are highly individual and specialized people. Each of them is marked by the kind of work he or she prefers and has done, which in time becomes an aspect of that individual's personality.

Margaret Mead, American Anthropologist (1901–1978)

5.16 Most people prefer to carry out the kinds of experiments that allow the scientist to feel that he is in full control of the situation rather than surrendering himself to the situation, as one must in studying human beings as they actually live.

Margaret Mead, American Anthropologist (1901–1978)

5.17 My happiness was complete when I learned that all my precious specimens had been saved. Large parts of my collections, many of my books, and all of my clothes had been stolen, but Early Man survived the disaster. [On the recovery of his invaluable collection of fossil skulls and teeth following his release from a Japanese prisoner of war camp.]

G.H.R. von Koenigswald, German Paleontologist (1902–)

5.18 The evidence indicates that woman is, on the whole, biologically superior to man.

Ashley Montagu, American Anthropologist/Biologist (1905–)

5.19 Anthropology provides no scientific basis for discrimination against any people on the ground of racial inferiority, religious affiliation, or linguistic heritage.

American Anthropological Association, Resolution (1938)

5.20 In the early seventies my mother [Mary Leakey] returned to Laetoli, a site in Tanzania that my parents had discovered in the thirties. In 1976, she discovered footprints of ancient hominids. These three-and-a-half-million-year-old footprints provide evidence of bipedality.

Richard Leakey, English Paleontologist (1944–)

5.21 In man, population structure reaches its greatest complexity. Mankind—the human species, *Homo sapiens*—is the most inclusive Mendelian population, one which inhabits nearly the whole globe.

Bruce Wallace and Theodosius Dobzhansky, American Geneticist and Russian/American Anthropologist (1959)

5.22 A traditional anthropological description is like a book of etiquette. What you get isn't so much the deep cultural wisdom as the cultural clichés, the wisdom of Polonius, conventions in the trivial rather than the informing sense. It may tell you the official rules, but it won't tell you how life is lived.

Renato Rosaldo, American Anthropologist (1986)

5.23 Whatever their degree of statistical rigor, social scientists build their knowledge on the sands of interpretation.

Louis A. Sass, American Anthropologist (1986)

5.24 Given the large quantities of meat and marrow available during hominid feeding events, it is likely that cooperative food sharing on a scale unknown among modern nonhuman primates occurred nearly two million years ago.
Bruce Bower, American Science Journalist (1987)

6.
Archaeology

Wait a thousand years and even the garbage

left behind by a vanished civilization

becomes precious to us.

ISAAC ASIMOV

6.1 They who built with granite, who set a hall inside their pyramid, and wrought beauty with their fine work . . . their altar stones also are empty as are those of the weary ones, the ones who die upon the embankment leaving no mourners.

Ancient Egyptian Text (circa 1700 B.C.)

6.2 O Mother Nut! Spread your winds above me like the imperishable stars!

Inscription on the coffin of Tutankhamen, Egyptian Pharoah (circa 1350 B.C.)

6.3 King Darayawaush gives notice thus:
You who in future days
Will see this inscription by order
Writ with hammer upon the cliff,
Who will see these human figures here—
Efface, destroy nothing.
Take care, so long as you have seed,
To leave them undisturbed.
[Inscribed on the cliffs of Bagistana.]

Darius I, Persian King (558 B.C.?–486 B.C.?)

6.4 Monuments are made for victories over foreigners: domestic troubles should be covered with the veil of sadness.

Julius Caesar, Roman Emperor/Tactician (104 B.C.–44 B.C.)

6.5 Not by marble inscribed with public records is the breath and life of goodly heroes continued after death.

Horace (Quintus Horatius Flaccus), Roman Poet (65 B.C.–8 B.C.)

6.6 To build up cities, an age is needed: but an hour destroys them.

Lucius Annaeus Seneca (the Younger), Roman Philosopher/Statesman/ Dramatist (4 B.C.?–65 A.D.)

6.7 Ye are like unto whited sepulchres, which indeed appear beautiful outward, but are within full of dead men's bones, and of all uncleanness.

***The Bible* (circa 325 A.D.)**

6.8 Babylon the great is fallen, is fallen.

***The Bible* (circa 325 A.D.)**

6.9 How poor remembrances are statues, tombs,
And other monuments that men erect
To princes, which remain closed rooms
Where but a few behold them.

John Florio, Italian/English Author (1553?–1625)

6.10 Gold once out of the earth is no more due unto it; what was unreasonably committed to the ground, is reasonably resumed from it; let monuments and rich fabrics, not riches, adorn men's ashes.

Thomas Browne, English Writer/Antiquarian (1605–1682)

6.11 The Egyptian mummies, which Cambyses or time hath spared, avarice now consumeth. Mummy is become merchandise, Mizraim cures wounds, and Pharaoh is sold for balsams.

Thomas Browne, English Writer/Antiquarian (1605–1682)

6.12 Papyra, throned upon the banks of Nile,
Spread her smooth leaf, and waved her silver style.
—The storied pyramid, the laurel'd bust,
The trophy'd arch had crumbled into dust;
The sacred symbol, and the epic song
(Unknown the character, forgot the tongue,)
With each unconquer'd chief, or sainted maid,
Sunk undistinguish'd in Oblivion's shade.
Sad o'er the scatter'd ruins Genius sigh'd,
And infant Arts but learn'd to lisp, and died.
Till to astonish'd realms Papyra taught
To paint in mystic colours Sound and Thought,
With Wisdom's voice to print the page sublime,
And mark in adamant the steps of Time.

Erasmus Darwin, English Physician/Poet (1731–1802)

6.13 But monuments themselves memorials need.

George Crabbe, English Poet (1754–1832)

6.14 What marvel is this? We begged you for drinkable springs,
O earth, and what is your lap sending forth?
Is there life in the deeps as well? A race yet unknown
Hiding under the lava? Are they who had fled returning?
Come and see, Greeks; Romans, come! Ancient Pompeii
Is found again, the city of Hercules rises!
Friedrich von Schiller, German Poet/Dramatist (1759–1805)

6.15 Soldiers! Forty centuries gaze down upon you!
[On seeing the Great Pyramids at Giza.]
Napoleon I (Napoleon Bonaparte), French Emperor/General (1769–1821)

6.16 May no rude hand deface it,
And its forlorn "Hic jacet."
William Wordsworth, English Poet (1770–1850)

6.17 While stands the Coliseum, Rome shall stand;
When falls the Coliseum, Rome shall fall;
And when Rome falls—the world.
Lord Byron (George Gordon), English Poet/Dramatist (1788–1824)

6.18 Let not a monument give you or me hopes,
Since not a pinch of dust remains of Cheops.
Lord Byron (George Gordon), English Poet/Dramatist (1788–1824)

6.19 My father, and the father of my father, pitched their tents here before me. . . . For twelve hundred years have the true believers—and, praise be to God! all true wisdom is with them alone—been settled in this country, and not one of them ever heard of a palace underground. Neither did they who went before them. But lo! here comes a Frank from many days' journey off, and he walks up to the very place, and he takes a stick . . . and makes a line here, and makes a line there. Here, says he, is the palace; there, says he, is the gate; and he shows us what has been all our lives beneath our feet, without our having known anything about it. Wonderful! Wonderful! Is it by books, is it by magic, is it by your prophets, that you have learnt wisdom?
[Spoken to Austin Layard, English archaeologist who discovered and excavated Nineveh and Nimrud, 1845–1861.]
Sheik Abd-er-Rahman, Sultan of Fez (circa 1850)

6.20 Men moralise among ruins.
Benjamin Disraeli, English Statesman/Writer (1804–1881)

6.21 Time will soon destroy the works of famous painters and sculptors, but the Indian arrowhead will balk his efforts and Eternity will have to come to his aid. They are not fossil bones, but, as it were, fossil thoughts, forever reminding me of the mind that shaped them. . . . Myriads of arrow-points lie sleeping in the skin of the revolving earth, while meteors revolve in space. The footprint, the mind-print of the oldest men.
Henry David Thoreau, American Writer/Naturalist (1817–1862)

6.22 Who shall doubt "the secret hid
Under Egypt's pyramid"
Was that the contractor did
Cheops out of several millions?
Rudyard Kipling, English Author/Poet (1865–1936)

6.23 There's a fascination frantic
In a ruin that's romantic.
Sir W.S. (William Schwenk) Gilbert, English Comic Dramatist (1836–1911)

6.24 Lord Canaveron: "Can you see anything?"
Howard Carter: "Yes, wonderful things."
[Lord Canaveron and Howard Carter's comments upon Carter's first entry to the tomb of Tutankhamen.]
Howard Carter and Lord Canaveron, English Archaeologist and English Philanthropist (1922)

6.25 If we human beings want to feel humility, there is no need to look at the starred infinity above. It suffices to turn your gaze upon the world cultures that existed thousands of years before us.
C.W. Ceram, English Author/Archaeologist (1915–1972)

7. Astronomy

Whatever else astronomy may or may not be who can doubt it to be the most beautiful of the sciences?

ISAAC ASIMOV

7.1 Canst thou bind the sweet influences of Pleiades, or loose the bans of Orion?

The Bible (circa 725 B.C.)

7.2 Strepsiades: But why do they look so fixedly on the ground?
Disciple of Socrates: They are seeking for what is below the ground.
Strepsiades: And what is their rump looking at in the heavens?
Disciple: It is studying astronomy on its own account.

Aristophanes, Greek Comic Dramatist (448 B.C.?–380 B.C.)

7.3 All of us Hellenes tell lies about those great gods, the sun and the moon. . . . We say that they, and diverse other stars, do not keep the same path, and we call them planets or wanderers. On the contrary, each of them moves in the same path—not in many paths, but in one only, which is circular, and the varieties are only apparent.

Plato, Greek Philosopher (427 B.C.?–347 B.C.?)

7.4 The contemplation of celestial things will make a man both speak and think more sublimely and magnificently when he descends to human affairs.

Marcus Tullius Cicero, Roman Orator/Statesman (106 B.C.–43 B.C.)

7.5 Happy the men who made the first essay,
And to celestial regions found the way!

No earthly vices clogg'd their purer souls,
That they could soar so high as touch the poles:
Sublime their thoughts and from pollution clear,
Bacchus and Venus held no revels there;
From vain ambition free; no love of war
Possess'd their minds, nor wranglings at the bar;
No glaring grandeur captivates their eyes,
For such see greater glory in the skies;
Thus these to heaven attain.

Ovid (Publius Ovidius Nasso), Roman Poet (43 B.C.–17 A.D.?)

7.6 At Lammas [1 August] of this year king Henry went oversea; and on the following day, while he lay asleep on board, the light of day was eclipsed over all lands, and the sun looked like a moon three nights old, and there were stars around it at midday.

Then men were greatly astonished and terrified, and said that some important event should follow upon this; and so it did, for in that very year the king died in Normandy the day after St. Andrew's day [30 November].

Anonymous, *The Anglo-Saxon Chronicle* (1135)

7.7 Let no one expect anything in the way of certainty from astronomy, since astronomy can offer us nothing certain, lest, if anyone take as true that which has been constructed for another use, he go away from this discipline a bigger fool than when he came to it.

Nicholas Copernicus, Polish Astronomer (1473–1543)

7.8 Astronomy is the most ancient of all the sciences, and has been the introducer of vast knowledge.

Martin Luther, German Theologian (1483–1546)

7.9 Now it is quite clear to me that there are no solid spheres in the heavens, and those that have been devised by the authors to save the appearances, exist only in the imagination, for the purpose of permitting the mind to conceive the motion which the heavenly bodies trace in their course.

Tycho Brahe, Danish Astronomer (1546–1601)

7.10 And teach me how
To name the bigger light, and how the less,
That burn by day and night.

William Shakespeare, English Dramatist/Poet (1564–1616)

7.11 This is the excellent foppery of the world, that when we are sick in fortune—often the surfeit of our own behaviour—we make guilty of our disasters the sun, the moon, and the stars: as if we were villains by necessity; fools by heavenly compulsion; knaves, thieves, and treachers by spherical predominance; drunkards, liars, and adulterers by an enforced obedience of planetary influence.

William Shakespeare, English Dramatist/Poet (1564–1616)

7.12 Not from the stars do I my judgement pluck,
And yet methinks I have astronomy;
But not to tell of good or evil luck,
Of plagues, of dearths, or seasons' quality;
Nor can I fortune to brief minutes tell,
Pointing to each his thunder, rain, and wind,
Or say with princes if it shall go well
By oft predict that I in heaven find;
But from thine eyes my knowledge I derive,
And, constant stars, in them I read such art
As truth and beauty shall together thrive
If from thyself to store thou wouldst convert:
Or else of thee this I prognosticate,
Thy end is truth's and beauty's doom and date.

William Shakespeare, English Dramatist/Poet (1564–1616)

7.13 Nature has given us astrology as an adjunct and ally to astronomy.

Johannes Kepler, German Astronomer/Mathematician (1571–1630)

7.14 I send herewith unto his Majesty the strangest piece of news (as I may justly call it) that he hath ever yet received from any part of the world; which is the annexed book (come abroad this very day) of the Mathematical Professor of Padua (the *Siderius Nuncius* by Galileo), who by the help of an optical instrument (which both enlargeth and approximateth the object) invented first in Flanders, and bettered by himself, hath discovered four new planets rolling about the sphere of Jupiter besides many other unknown fixed stars; likewise the true cause of the Via Lactae . . . and lastely, that the moon is not spherical, but endued with many prominences. . . . So, as upon the whole subject he hath first overthrown all former astronomy—for we must have a new sphere to save appearances—and next all astrology. For the first of the new planets must vary the judicial part, and why may there not yet be more?

Henry Wooten, Architect/Military Engineer (circa 1610)

7.15 Against filling the heavens with fluid mediums, unless they be exceeding rare, a great objection arises from the regular and very lasting motions of the planets and comets in all manner of courses through the heavens.

Isaac Newton, English Physicist/Mathematician (1642–1727)

7.16 Astronomy was the daughter of idleness.

Bernard Le Bovier Sieur de Fontenelle, French Philosopher (1657–1757)

7.17 When a man spends his life among the stars and planets, or lays out a twelvemonth on the spots of the sun, however noble his speculations may be, they are very apt to fall into burlesque.

Joseph Addison, English Essayist/Poet/Politician (1672–1719)

7.18 Astrology was much in vogue during the middle ages, and became the parent of modern astronomy, as alchemy did of chemistry.

Noah Webster, American Author/Lexicographer/Philologist (1758–1843)

7.19 It is not for us to say whether Inspiration revealed to the Psalmist the wonders of astronomy. But even though the mind be a perfect stranger to the science of these enlightened times, the heavens present a great and an elevating spectacle—an immense concave reposing on the circular boundary of the world, and the innumerable lights which are suspended from on high, moving with solemn regularity along its surface.

Thomas Chalmers, Scottish Theologian/Author (1780–1847)

7.20 We have something more than the mere magnitude of the planets to allege, in favor of the idea that they are inhabited. . . . though this mighty earth, with all its myriads of people, were to sink into annihilation, there are some worlds where an event so awful to us would be unnoticed and unknown, and others where it would be nothing more than the disappearance of a little star which had ceased from its twinkling.

Thomas Chalmers, Scottish Theologian/Author (1780–1847)

7.21 In her starry shade
Of dim and solitary loveliness,
I learn the language of another world.

Lord Byron (George Gordon), English Poet/Dramatist (1788–1824)

7.22 In fields of air he writes his name,
And treads the chambers of the sky;
He reads the stars, and grasps the flame
That quivers in the realms on high.

Charles Sprague, American Poet/Essayist/Banker (1791–1875)

7.23 Astronomy is the science of the harmony of infinite expanse.

Lord John Russell, English Statesman/Author (1792–1878)

7.24 Astronomy is one of the sublimest fields of human investigation. The mind that grasps its facts and principles receives something of the enlargement and grandeur belonging to the science itself. It is a quickener of devotion.

Horace Mann, American Educator/Politician (1796–1859)

7.25 When I, sitting, heard the astronomer, where he lectured with
such applause in the lectureroom,
How soon, unaccountable, I became tired and sick;
Till rising and gliding out, I wander'd off by myself,
In the mystical moist night-air, and from time to time,
Look'd up in perfect silence at the stars.

Walt Whitman, American Poet (1819–1892)

7.26 An undevout astronomer is mad.

Charles Augustus Young, American Astronomer (1834–1908)

7.27 Night after night, among the gabled roofs,
Climbing and creeping through a world unknown

Save to the roosting stork, he learned to find
The constellations, Cassiopeia's throne,
The Plough still pointing to the Polar Star,
The movements of the planets, hours and hours,
And wondered at the mystery of it all.
Alfred Noyes, English Poet (1880–1958)

7.28 The oldest picture book in our possession is the Midnight Sky.
E. Walter Maunder, English Writer (1900)

7.29 In the streets of a modern city the night sky is invisible; in rural districts, we move in cars with bright headlights. We have blotted out the heavens, and only a few scientists remain aware of stars and planets, meteorites and comets.
Bertrand Russell, English Philosopher/Mathematician (1952)

7.30 In the path of a total eclipse one has the privilege of sharing the perfect alignment of earth, moon, and sun. For a brief interval out of time, the three bodies are frozen in majestic union; then they go on their separate, complicated courses. To participate in that moment of uncanny equilibrium is to have one's faith strengthened in the possibility of equilibrium and to experience the paradox that balance and stillness are to be found at the heart of all change.
Andrew Weil, American Ethnopharmacologist (1980)

8.
The Atom

Circumstantial evidence can be overwhelming. We have never seen an atom, but we nevertheless know that it must exist.

ISAAC ASIMOV

8.1 In reality, nothing but atoms and the void.
Democritus of Adbera, Greek Philosopher (460 B.C.–371 B.C.)

8.2 Globed from the atoms falling slow or swift
I see the suns, I see the systems live
Their forms; and even the systems and the suns
Shall go back slowly to the eternal drift.
Lucretius, Roman Poet/Philosopher (96 B.C.?–55 B.C.?)

8.3 If the atoms have by chance formed so many sorts of figures, why did it never fall out that they made a house or a shoe? Why at the same rate should we not believe that an infinite number of Greek letters strown all over a certain place might possibly fall into the contexture of the Iliad?
Michel de Montaigne, French Essayist (1533–1592)

8.4 I will paint for man not only the visible universe, but all that he can conceive of nature's immensity in the womb of an atom.
Blaise Pascal, French Mathematician/Philosopher/Author (1623–1662)

8.5 We shall never get people whose time is money to take much interest in atoms.
Samuel Butler, English Novelist (1835–1902)

8.6 The conception of the atom stems from the concepts of subject and substance: there has to be "something" to account for any action. The atom is the last descendant of the concept of the soul.

Friedrich Nietzsche, German Philosopher (1844–1900)

8.7 And even your atom, my dear mechanists and physicists—how much error, how much rudimentary psychology is still residual in your atom!

Friedrich Nietzsche, German Philosopher (1844–1900)

8.8 The wave phenomenon forms the real "body" of the atom. It replaces the individual punctiform electrons, which in Bohr's model swarm around the nucleus.

Erwin Schrödinger, Austrian Physicist (1887–1961)

8.9 Now I know what the atom looks like!

Ernest Rutherford, British Physicist (1911)

8.10 In 1914. . . . the atomic weight of lead of radioactive origin—lead obtained by Boltwood from a uraninite from North Carolina—was found to be substantially smaller than that of ordinary lead—a thrilling discovery to chemists, for never before in the history of science had the atomic weight of a chemical element been known to vary.

Adolph Knopf, American Geologist (1941)

8.11 Today we not only have no perfect model [of the atom] but we know that it is of no use to search for one.

Sir James Hopwood Jeans, English Mathematician/Astronomer (1942)

8.12 The Atomic Age began at exactly 5:30 Mountain War Time on the morning of July 15, 1945, on a stretch of semi-desert land about 50 airline miles from Alamogordo, New Mexico. And just at that instance there rose from the bowels of the earth a light not of this world, the light of many suns in one.

William L. Laurence, American Journalist (1888–1977)

8.13 Nature is neutral. Man has wrested from nature the power to make the world a desert or to make the deserts bloom. There is no evil in the atom; only in men's souls.

Adlai Ewing Stevenson, American Politician (1900–1965)

8.14 The mathematically formulated laws of quantum theory show clearly that our ordinary intuitive concepts cannot be unambiguously applied to the smallest particles. All the words or concepts we use to describe ordinary physical objects, such as position, velocity, color, size, and so on, become indefinite and problematic if we try to use them of elementary particles.

Werner Karl Heisenberg, German Physicist (1958)

8.15 [We should] abandon all attempts to construct perceptual models of atomic processes.

Werner Karl Heisenberg, German Physicist (1958)

8.16 In the years since man unlocked the power stored up within the atom, the world has made progress, halting but effective, toward bringing that power under human control. The challenge may be our salvation. As we begin to master the destructive potentialities of modern science, we move toward a new era in which science can fulfill its creative promise and help bring into existence the happiest society the world has ever known.

John F. Kennedy, American President (1960)

8.17 An elementary particle is not an independently existing, unanalyzable entity. It is, in essence, a set of relationships that reach outward to other things.

Henry Stapp, American Physicist (1972)

9. Biochemistry

The significant chemicals of living tissue are rickety and unstable, which is exactly what is needed for life.

ISAAC ASIMOV

9.1 Research gave the unexpected result that, by combination of cyanic acid with ammonia, urea is formed. A noteworthy fact since it furnishes an example of the artificial production of an organic—indeed, a so-called *animal*—substance from inorganic materials!

Friedrich Wöhler, German Chemist (1800–1882)

9.2 The animal frame, though destined to fulfill so many other ends, is as a machine more perfect than the best contrived steam-engine—that is, is capable of more work with the same expenditure of fuel.

James Prescott Joule, English Physicist (1818–1889)

9.3 It [the Euglena] is a perfect laboratory in itself, and it will act and react upon the water and the matters contained therein; converting them into new compounds resembling its own substance, and at the same time giving up portions of its own substance which have become effete.

Thomas Henry Huxley, English Biologist/Evolutionist (1825–1895)

9.4 If it be urged that the action of the potato is chemical and mechanical only, and that it is due to the chemical and mechanical effects of light and heat, the answer would seem to be in an inquiry whether every sensation is not chemical and mechanical in its operation? Whether those things which we deem most purely spiritual are anything but disturbances of equilibrium in an infinite series of levers, beginning with those that are too small for microscopic detection, and going up to the human arm and the applicances which it makes use of? Whether there be not a molecular action of thought, whence a dynamical theory of the passions shall be deducible?

Samuel Butler, English Novelist (1835–1902)

9.5 The activity of a neuron is an "all-or-none" process.

Warren McCullock, American Logician/Neurophysiologist (1899–1969)

9.6 Like other branches of biology, biochemistry can be considered as divisible into two main branches. There is first of all the problem of the chemical composition of living stuff, corresponding to the structural, morphological, or "static" approach which, in the end, amounts to a very specialized kind of organic chemistry. It is the second, the physiological and essentially dynamic approach, that is most characteristic of biochemistry today: the primary question has shifted from "what is it made of?" to "how does it work?"

Ernest Baldwin, American Biochemist (1955)

9.7 If, as a chemist, I see a flower, I know all that is involved in synthesizing a flower's elements. And I know that even the fact that it exists is not something that is natural. It is a miracle.

Albert Hofmann, Swiss Chemist (1984)

9.8 The special vital forces that distinguish living things from the nonliving are emergent, holistic properties, not properties of their physiochemical components. Nor can they be explained in mechanistic terms.

Roger Wolcott Sperry, American Neurophysiologist/Psychobiologist (1984)

10. Biology

What can be more important than the science of life to any intelligent being who has the good fortune to be alive?

ISAAC ASIMOV

10.1 The Material World is only the Shell of the Universe: The World of Life are its Inhabitants.

Joseph Addison, English Essayist/Poet/Politician (1672–1719)

10.2 We must ascribe to all cells an independent vitality; that is, such combinations of molecules as occur in any single cell are capable of setting free the power by which it is enabled to take up fresh molecules.

Theodor Schwann, German Physiologist/Biologist (1810–1882)

10.3 Living things have no inertia, and tend to no equilibrium.

Thomas Henry Huxley, English Biologist/Evolutionist (1825–1895)

10.4 Teleology is a lady without whom no biologist can live; yet he is ashamed to show himself in public with her.

Franklin von Bruecke, German Biologist (1845–1916)

10.5 The living being is stable. It must be so in order not to be destroyed, dissolved, or disintegrated by the colossal forces, often adverse, which surround it. By apparent contradiction it maintains its stability only if it is excitable and capable of modifying itself according to external stimuli and adjusting its response to the stimulation.

Charles Robert Richet, French Physiologist (1850–1935)

10.6 Biology cannot go far in its subject without being met by mind.
Sir Charles Scott Sherrington, English Physiologist (1857–1952)

10.7 Every living thing is a sort of imperialist, seeking to transform as much as possible of its environment into itself and its seed.
Bertrand Russell, English Philosopher/Mathematician (1872–1970)

10.8 The cause of every need of a living being is also the cause of the satisfaction of the need.
Eduard Friedrich Wilhelm Pfluger, German Physiologist (circa 1877)

10.9 For the biologist there are no classes—only individuals.
Jean Rostand, French Biologist (1894–1977)

10.10 [Biology] is the least self-centered, the least narcissistic of the sciences—the one that, by taking us out of ourselves, leads us to re-establish the link with nature and to shake ourselves free from our spiritual isolation.
Jean Rostand, French Biologist (1894–1977)

10.11 The role of biology today, like the role of every other science, is simply to describe, and when it explains it does not mean that it arrives at finality; it only means that some descriptions are so charged with significance that they expose the relationship of cause and effect.
Donald Culross Peattie, American Botanist/Author (1898–)

10.12 Life-cycle completion is indeed the master law governing all the activities of the organism, to which other laws of smaller scope, such as the law of need-satisfaction, are subordinate.
E.S. Russell, English Scientist/Philosopher (1950)

10.13 I . . . object to dividing the study of living processes into botany, zoology, and microbiology because by any such arrangement, the interrelations within the biological community get lost. Corals cannot be studied without reference to the algae that live with them; flowering plants without the insects that pollinate them; grasslands without the grazing mammals.
Marston Bates, American Zoologist/Science Journalist (1960)

10.14 For as long as I can remember, I have been fascinated by anything that lives; one is either born a naturalist or one never becomes one.
Jean Delacour, British Naturalist/Writer (1966)

10.15 Society increasingly has neglected the substructure of biology, to its own peril.
Edward O. Wilson, American Entomologist/Sociobiologist (1984)

10.16 We need to explore the possiblity that homosexual bonding may be a biological mechanism.
Edward O. Wilson, American Entomologist/Sociobiologist (1984)

10.17 Learning and memory are just as much biology as a process of DNA replication.

Anke A. Ehrhardt, American Psychiatrist/Educator (1986)

10.18 Examination of the folds and ridges of the human brain . . . reveals an individuality unguessed at by earlier investigators. No two genetic programs or environments or life experiences are ever precisely alike or ever could be. Provide small fluctuations in the environment or diet . . . and that brain takes a different turn. . . . Experience, fate, environment, and chance exert their effects on the fetal brain, mold it, shape it, and indeed sculpt it toward a pattern that, however many brains may be examined until doomsday, will be found in that brain and that brain alone.

Richard M. Restak, American Writer/Biologist (1986)

10.19 The division between life and nonlife is perhaps an artificial one.

Cyril Ponnamperuma, American Chemist (1986)

10.20 Plasma seems to have the kinds of properties one would like for life. It's somewhat like liquid water—unpredictable and thus able to behave in an enormously complex fashion. It could probably carry as much information as DNA does. It has at least the potential for organizing itself in interesting ways.

Freeman Dyson, American Mathematician (1986)

11.
Botany

It's humbling to think that all animals,

including human beings, are parasites

of the plant world.

ISAAC ASIMOV

11.1 Consider the lilies of the field, how they grow; they toil not, neither do they spin.

The Bible (circa 325 A.D.)

11.2 Flowers often grow more beautifully on dung-hills than in gardens that look beautifully kept.

Saint Francis of Sales, French Roman Catholic Prelate/Author (1567–1622)

11.3 The seed of a tree has the nature of a branch or twig or bud. While it grows upon the tree it is a part of the tree: but if separated and set in the earth to be better nourished, the embryo or young tree contained in it takes root and grows into a new tree.

Isaac Newton, English Physicist/Mathematician (1642–1727)

11.4 A practical botanist will distinguish at the first glance the plant of the different quarters of the globe and yet will be at a loss to tell by what marks he detects them.

Carl Linnaeus, Swedish Botanist/Naturalist (1707–1778)

11.5 Tantus amor florum.
[So great is the love of flowers.]

Gilbert White, English Naturalist/Ecologist (1720–1793)

11.6 Plants, instead of affecting the air in the same manner as animal respiration, reverse the effect of breathing and tend to keep the atmosphere sweet and wholesome.

Joseph Priestly, English Chemist (1733–1804)

11.7 The greatest service which can be rendered any country is to add an useful plant to its culture; especially, a bread grain; next in value to bread is oil.

Thomas Jefferson, American President/Author (1743–1826)

11.8 The archetypal plant will be the strangest growth the world has ever seen, and nature herself shall envy me for it. With such a model, and with the key to it in one's hands, one will be able to contrive an infinite variety of plants. They will be strictly logical plants—in other words, even though they may not actually exist, they could exist.

Johann Wolfgang von Goethe, German Poet/Dramatist/Novelist (1749–1832)

11.9 Where flowers degenerate man cannot live.

Napoleon I (Napoleon Bonaparte), French Emperor/General (1769–1821)

11.10 And all their botany is Latin names.

Ralph Waldo Emerson, American Essayist/Philosopher/Poet (1803–1882)

11.11 Weed—a plant whose virtues have not yet been discovered.

Ralph Waldo Emerson, American Essayist/Philosopher/Poet (1803–1882)

11.12 [The root cap of a plant], having the power of directing the movements of the adjoining parts, acts like the brain of one of the lower animals; the brain being seated within the anterior end of the body, receiving impressions from the sense-organs, and directing the several movements.

Charles Robert Darwin, English Naturalist/Evolutionist (1809–1882)

11.13 The little beggars are doing just what I don't want them to. [About the plants in his garden.]

Charles Robert Darwin, English Naturalist/Evolutionist (1809–1882)

11.14 When you find that flowers will not endure a certain atmosphere, it is a very significant hint to the human creature to remove out of that neighborhood.

Henry Mayhew, English Journalist (1812–1887)

11.15 Flowers are words,
Which even a babe may understand.

Arthur Cleveland Coxe, American Clergyman (1818–1896)

11.16 History records the names of royal bastards, but it cannot tell us the origin of wheat.

Jean Henri Fabre, French Entomologist/Naturalist (1823–1915)

11.17 I pull a flower from the woods,
A monster with a glass
Computes the stamens in a breath,
And has her in a class.

Emily Dickinson, American Poet (1830–1886)

11.18 It is not so much that the cells make the plant; it is rather that the plant makes the cells.

Heinrich Anton de Bary, German Botanist/Physician (1831–1888)

11.19 There is a kind of plant that eats organic food with its flowers: when a fly settles upon the blossom, the petals close upon it and hold it fast till the plant has absorbed the insect into its system; but they will close on nothing but what is good to eat; of a drop of rain or a piece of stick they will take no notice. Curious! that so unconscious a thing should have such a keen eye to its own interest.

Samuel Butler, English Novelist (1835–1902)

11.20 Beware of old Linnaeus,
The Man of the Linden-tree,
So beautiful, bright and early
He brushed away the dews
He found the wicked wild-flowers
All courting there in twos.

Alfred Noyes, English Poet (1880–1958)

11.21 We once had a lily here that bore *108* flowers on one stalk: it was photographed naturally for all the gardening papers. The bees came from miles and miles, and there were the most disgraceful Bacchanalian scenes: bees hardly able to find their way home.

Dame Edith Sitwell, English Poet/Author (1887–1964)

11.22 Flowers are restful to look at. They have neither emotions nor conflicts.

Sigmund Freud, Austrian Psychiatrist/Psychoanalyst (1856–1939)

11.23 I once saw a botanist most tenderly replace a plant which he had inadvertently uprooted, though we were on a bleak hillside in Tibet, where no human being was likely to see the flower again.

Sir Francis Edward Younghusband, English Explorer/Mystic/Author/ Soldier (1863–1942)

11.24 One in ten plant species contains anticancer substances of variable potency, but relatively few have been bioassayed.

Edward O. Wilson, American Entomologist/Sociobiologist (1985)

11.25 As important as the techniques of molecular biology are for the study of plant hormones, it is essential that we not lose sight of the integrated function of the plant as a whole.

Michael L. Evans, American Botanist (1986)

11.26 A very general rule of thumb is that every increment of 1,000 feet above sea level is equivalent to 600 miles in a south-north direction. By ascending a mountain one encounters life zones in the same sequence that one observes driving hundreds of miles from south to north.

Robert H. Mohlenbrock, American Botanist/Educator (1986)

12. Chemistry

For many centuries chemists labored to change lead into precious gold, and eventually found that precious uranium turned to lead without any human effort at all.

ISAAC ASIMOV

12.1 The various elements had different places before they were arranged so as to form the universe. At first, they were all without reason and measure. But when the world began to get into order, fire and water and earth and air had only certain faint traces of themselves, and were altogether such as everything might be expected in the absence of God; this, I say, was their nature at that time, and God fashioned them by form and number.

Plato, Greek Philosopher (427 B.C.?–347 B.C.?)

12.2 I do not find that anyone has doubted that there are four elements. The highest of these is supposed to be fire, and hence proceed the eyes of so many glittering stars. The next is that spirit, which both the Greeks and ourselves call by the same name, air. It is by the force of this vital principle, pervading all things and mingling with all, that the earth, together with the fourth element, water, is balanced in the middle of space.

Pliny the Elder (Gaius Plinius Secundus), Roman Naturalist/Historian (23–79)

12.3 We say in general that the material of all stone is either some form of Earth or some form of Water. For one or the other of these elements predominates in stones; and even in stones in which some form of Water seems to predominate, something of Earth is also important. Evidence of this is that nearly all kinds of stones sink in water.

Albertus Magnus, German Clergyman/Theologian/Scholar (1200?–1280)

12.4 The science of alchemy I like very well. I like it not only for the profits it brings in melting metals, in decocting, preparing, extracting, and distilling herbs, roots; I like it also for the sake of the allegory and secret signification, which is exceedingly fine, touching the resurrection of the dead at the last day.

Martin Luther, German Theologian (1483–1546)

12.5 On consideration and by the advice of learned men, I thought it improper to unfold the secrets of the art (alchemy) to the vulgar, as few persons are capable of using its mysteries to advantage and without detriment.

Tycho Brahe, Danish Astronomer (1546–1601)

12.6 Alchemy may be compared to the man who told his sons of gold buried somewhere in his vineyard, where they by digging found no gold, but by turning up the mould about the roots of their vines, procured a plentiful vintage. So the search and endeavors to make gold have brought many useful inventions and instructive experiments to light.

Francis Bacon, English Philosopher/Essayist/Statesman (1561–1626)

12.7 The world hath been much abused by the opinion of making gold; the work itself I judge to be possible; but the means hitherto propounded are (in the practice) full of error.

Francis Bacon, English Philosopher/Essayist/Statesman (1561–1626)

12.8 An Alchimist, That's all too much. Chimist you might him call And I think it were true, and leave out Al.

Sir John Harington, English Writer/Translator (1561–1612)

12.9 I mean by elements certain primitive and simple, or perfectly unmingled bodies which not being made of any other bodies . . . are the ingredients of which all those called perfectly mixed bodies are immediately compounded, and into which they are ultimately resolved.

Robert Boyle, British Physicist/Chemist (1627–1691)

12.10 I have always looked upon alchemy in natural philosophy, to be like over enthusiasm in divinity, and to have troubled the world much to the same purpose.

William Temple, English Statesman/Essayist (1628–1699)

12.11 Chemistry, an art whereby sensible bodies contained in vessels . . . are so changed, by means of certain instruments, and principally fire, that their several powers and virtues are thereby discovered, with a view to philosophy or medicine.

Samuel Johnson, English Lexicographer/Poet/Critic (1709–1784)

12.12 Thou, Chemistry, do penetrate
With vision keen the bowels of earth,
Reveal what treasure Russia hides there. . . .
Mikhail Vasilievich Lomonosov, Russian Chemist/Physicist/Poet (1711–1765)

12.13 *Nymphs!* you disjoin, unite, condense, expand,
And give new wonders to the Chemist's hand;
On tepid clouds of rising steam aspire,
Or fix in sulphur all its solid fire;
With boundless spring elastic airs unfold,
Or fill the fine vacuities of gold
With sudden flash vitrescent sparks reveal,
By fierce collision from the flint and steel. . . .
Erasmus Darwin, English Physician/Poet (1731–1802)

12.14 I have procured air [oxygen] . . . between five and six times as good as the best common air that I have ever met with.
Joseph Priestly, English Chemist (1733–1804)

12.15 Neither cookery nor chemistry [has] been able to make milk out of grass.
William Paley, English Educator/Theologian (1743–1805)

12.16 The ultimate particles of all homogeneous bodies are perfectly alike in weight, figure, & c[hemistry].
John Dalton, English Physicist/Chemist (1766–1844)

12.17 Chemical analysis and synthesis go no farther than to the separation of particles one from another, and to their reunion. No new creation or destruction of matter is within the reach of chemical agency. We might as well attempt to introduce a new planet into the solar system, or to annihilate one already in existence, as to create or destroy a particle of hydrogen.
John Dalton, English Physicist/Chemist (1766–1844)

12.18 Organic chemistry just now is enough to drive one mad. It gives me the impression of a primeval tropical forest full of the most remarkable things, a monstrous and boundless thicket with no way of escape, and into which one may well dread to enter.
Baron Jöns Jakob Berzelius, Swedish Chemist (1779–1848)

12.19 The ancient teachers of this science [chemistry] said he, "Promised impossibilities and performed nothing." The modern masters promise very little; they know that metals cannot be transmuted and that the elixir of life is a chimera. But these philosophends [sic] seem only made to dabble in dirt, and their eyes to pore over the microscope or crucible, have indeed performed miracles. They penetrate into the recesses of nature and show how she works in her hiding-places. They ascend into the heavens; they have discovered how the blood circulates, and the nature of the air we breathe. They can command the thunders of heaven, mimic the earthquake, and even mock the invisible world with its own shadows.

Mary Wollstonecraft Shelley, English Novelist (1797–1851)

12.20 Chemistry is that branch of natural philosophy in which the greatest improvements have been and may be made; it is on that account that I have made it my peculiar study; but at the same time, I have not neglected the other branches of science. A man would make but a very sorry chemist if he attended to that department of human knowledge alone. If your wish is to become really a man of science and not merely a petty experimentalist, I should advise you to apply to every branch of natural philosophy, including mathematics.

Mary Wollstonecraft Shelley, English Novelist (1797–1851)

12.21 A chemical compound once formed would persist for ever, if no alteration took place in the surrounding conditions.

Thomas Henry Huxley, English Biologist/Evolutionist (1825–1895)

12.22 The Chemical conviction
That Nought be lost
Enable in Disaster
My fractured Trust—
The Faces of the Atoms
If I shall see
How more the Finished Creatures
Departed Me!

Emily Dickinson, American Poet (1830–1886)

12.23 For the first time I saw a medley of haphazard facts fall into line and order. All the jumbles and recipes and Hotchpotch of the inorganic chemistry of my boyhood seemed to fit themselves into the scheme before my eyes—as though one were standing beside a jungle and it suddenly transformed itself into a Dutch garden. 'But it's true,' I said to myself. 'It's very beautiful. And it's true.'

Baron C.P. (Charles Percy) Snow, English Novelist/Physicist (1934)

13. Comets and Meteors

How bright and beautiful a comet is as it flies past our planet — provided it does *fly past it.*

ISAAC ASIMOV

13.1 Why, then, are we surprised that comets, such a rare spectacle in the universe, are not known, when their return is at vast intervals?. . . . The time will come when diligent research over very long periods will bring to light things which now lie hidden. . . . Someday there will be a man who will show in what regions comets have their orbit, why they travel so remote from other celestial bodies, how large they are and what sort they are.

Lucius Annaeus Seneca (the Younger), Roman Philosopher/Statesman/ Dramatist (4 B.C.?–65 A.D.)

13.2 In the city of Tours on 31 January in the eighth year of the reign of King Childebert [583 A.D.] this day being Sunday, the bell had just rung for matins. The people had got up and were on their way to church. The sky was overcast and it was raining. Suddenly a great ball of fire fell from the sky and moved some considerable distance through the air, shining so brightly that visibility was as clear as at high noon. Then it disappeared once more behind a cloud and darkness fell again. The rivers rose much higher than usual.

Gregory of Tours, Frankish Ecclesiastic/Historian (538–594)

13.3 At that time, throughout all England, a portent such as men had never seen before was seen in the heavens. Some declared that the star was a comet, which some call "the long-haired star": it first appeared on the eve of the festival of *Letania Maior,* that is on 24 April, and shone every night for a week.

Anonymous, *The Anglo-Saxon Chronicle* (1066)

13.4 In the first week of Lent, on the Friday, 16 February, a strange star appeared in the evening, and for a long time afterwards was seen shining for a while each evening. The star made its appearance in the south-west, and seemed to be small and dark, but the light that shone from it was very bright, and appeared like an enormous beam of light shining north-east; and one evening it seemed as if the beam were flashing in the opposite direction towards the star. Some said that they had seen other unknown stars about this time, but we cannot speak about these without reservation, because we did not ourselves see them.

Anonymous, *The Anglo-Saxon Chronicle* (1105)

13.5 A comet is sublimated fire assimilated to the nature of one of the seven planets.

Robert Grosseteste, English Bishop/Educator (1168–1253)

13.6 No further that yeir, bot a strang meteor, quhilk was hard and sein in the seventh day of December. About ane houre befoir the sone rose, the moone schyneing, their wes at ane instant sein gryt inflamatiounes of fyre-flauchtis in the eisterne hemisphere, and suddentlie thaireftir wes hard a gryt crack, as of a gryt cannoun, and sensiblie markit a gryt glob or bullat, fyrrie-cullorit, with a mychtie quhissilling noyse, flieing from the north-eist to the south-west, quhilk left begin a blew traine and draught in the air, most lyk ane serpent in mony faulds and likit wimples; the heid quahairof breathing out flames and smooke, as it wald directlie invaid the moone, and swallowit hir up; but immediatlie the sone fyseing, faire and pleasant, abolischit all.

James Melville, Scottish Theologian (1556–1613)

13.7 Who vagrant transitory comets sees,
Wonders because they're rare; but a new star
Whose motion with the firmament agrees,
Is miracle; for there no new things are.

John Donne, English Poet/Essayist/Clergyman (1572–1631)

13.8 This hairy meteor did announce
The fall of sceptres and of crowns.

Samuel Butler, English Satirist (1612–1680)

13.9 Aristotle's opinion . . . that comets were nothing else than sublunary vapors or airy meteors . . . prevailed so far amongst the Greeks, that this sublimest part of astronomy lay altogether neglected; since none could think it worthwhile to observe, and to give an account of the wandering and uncertain paths of vapours floating in the Ether.

Edmond Halley, English Astronomer (1656–1742)

13.10 If I were a comet, I should consider the men of our present age a degenerate breed.

Bertrand Russell, English Philosopher/Mathematician (1872–1970)

13.11 About 0.1 percent of all material which has ever fallen on earth is organic. By comparison, if we measure the total weight of all organic matter on earth against the mass of the planet itself, only 0.0000001 percent is of living origin. This means that meteors are coming from somewhere that is a million times more organic than earth itself.

Lyall Watson, American Biologist (1979)

13.12 It is possible that organic molecules may take up as much as thirty percent of the mass of a comet—a concentration greater than known for any other situation in the universe, including the surface of the earth.

Lyall Watson, American Biologist (1979)

13.13 Of the 10,000 or so meteorites that have been collected and analyzed, eight are particularly unusual. They are so unusual, in fact, that since 1979 some investigators have thought they might have originated not in asteroids, as most meteorites did, but on the surface of Mars.

Lawrence M. Krauss, American Physicist/Astronomer (1986)

13.14 We cannot help but wonder, as Halley's leaves our sky to begin another of its long cycles, what kind of world it will find when it returns in three quarters of a century. No serious person puts much credence in the old belief in comets as "evil stars." Yet no one can deny Halley's link with so many great events in human history. It is as though the comet were a mirror, reminding us of what humanity was like and was doing each time it returned. If the history it mirrors in its thirty visits is all too often disturbing, that is the way human affairs usually were! The power to change the history is in us, not in the comet.

Thomas D. Nicholson, American Natural History Museum Director (1986)

13.15 A series of dark spots . . . have been appearing in ultraviolet images of earth's atmosphere taken by the Dynamics Explorer 1 satellite . . . ever since the craft was launched in 1981. Several of the spots show up in virtually every one of the more than 10,000 images so far amassed by (Louis) Frank and colleagues John Sigwarth and John Craven. . . .

The furor arises from Frank *et al.*'s proffered explanation: that the spots represent reductions in the atmosphere's ultraviolet brightness, triggered by the water from vast numbers of small, previously unsuspected comets. . . . so many that their mass would add up to the equivalent, in water, of earth's entire atmosphere every 5 million years.

Jonathan Eberhart, American Science Journalist (1986)

13.16 You can be absolutely sure that if you have a handful of particles from Greenland material, you've got pieces of Halley's comet, pieces of Kohoutek, and pieces of Swift-Tuttle.

Donald Brownlee, American Astronomer (1986)

13.17 Over very long time scales, when the perturbing influences of both Jupiter and Saturn are taken into account, the seemingly regular orbits of asteroids that stray into the Kirwood gaps turn chaotic.

For millions of years . . . such an orbit seems predictable. Then the path grows increasingly eccentric until it begins to cross the orbit of Mars and then the Earth. Collisions or close encounters with those planets are inevitable.

James Gleick, American Science Journalist (1987)

14.
Computer Science

Computers are better than we are at arithmetic, not because computers are so good at it, but because we are so bad at it.

ISAAC ASIMOV

14.1 It is unworthy of excellent men to lose hours like slaves in the labor of calculation which could safely be relegated to anyone else if machines were used.

Baron Gottfried Wilhelm von Leibnitz, German Mathematician/ Physicist/Philosopher (1646–1716)

14.2 The whole of the developments and operations of analysis are now capable of being executed by machinery. . . . As soon as an Analytical Engine exists, it will necessarily guide the future course of science.

Charles Babbage, English Inventor/Mathematician (1792–1871)

14.3 We may say most aptly that the Analytical Engine *weaves algebraic patterns* just as the Jacquard-loom weaves flowers and leaves.
[Comparing Charles Babbage's second proposed computing machine with the automated punch-card loom of J.M. Jacquard.]

Ada Augusta Lovelace, English Author/Science Historian (1815–1850)

14.4 The Analytical Engine [of Charles Babbage] has no pretensions whatever to originate anything. It can do whatever we *know how to order* it to perform. It can *follow* analysis; but it has no power of *anticipating* any analytical relations or truths. Its province is to assist us in making *available* what we are already acquainted.

Ada Augusta Lovelace, English Author/Science Historian (1815–1850)

14.5 But who can say that the vapour engine has not a kind of consciousness? Where does consciousness begin, and where end? Who can draw the line? Who can draw any line? Is not everything interwoven with everything? Is not machinery linked with animal life in an infinite variety of ways?

Samuel Butler, English Novelist (1835–1902)

14.6 It is by no means hopeless to expect to make a machine for really very difficult mathematical problems. But you would have to proceed step-by-step. I think electricity would be the best thing to rely on.

Charles Sanders Peirce, American Philosopher/Logician (1839–1914)

14.7 The computer is the most significant of human inventions because it complements the human brain in precisely the two ways which limit the brain—slowness and boredom. . . . It has added speed to the complexity of which would never be attempted because of the tedium involved.

Colin Pittendrigh, English Biologist (1918–)

14.8 I propose to consider the question, "Can Machines Think?"

Alan Mathison Turing, British Computer Scientist/Mathematician (1936)

14.9 If Babbage had lived seventy-five years later I would have been out of a job.

[Aiken created the Mark I computer.]

Howard Aiken, American Computer Scientist (1940)

14.10 The original question, "Can machines think?," I believe too meaningless to deserve discussion. Nevertheless I believe that at the end of the century the use of words and general educated opinion will have altered so much that one will be able to speak of machines thinking without expecting to be contradicted.

Alan Mathison Turing, British Computer Scientist/Mathematician (1950)

14.11 His [Alan Turing's] high-pitched voice already stood out above the general murmur of well-behaved junior executives grooming themselves for promotion within the Bell corporation. Then he was suddenly heard to say: "No, I'm not interested in developing a *powerful* brain. All I'm after is just a *mediocre* brain, something like the President of the American Telephone and Telegraph Company."

Andrew Hodges, English Mathematician (1949–)

14.12 It will be interesting to see if these machines play in the next decade the part of the cyclotrons and high voltage generators of the "thirties.". . . There is much to be said for digital computers as research projects for the time being; they are not so expensive as cyclotrons, they are much less messy, they are even more incomprehensible, and perhaps before very long they too will have to be taken over by the big firms.

B.V. Bowden, English Computer Scientist (1953)

14.13 All kinds of mistakes [in the computers] are turning up. You'd be surprised to know the number of doctors who claim they are treating pregnant men.

Anonymous, Canadian Health Insurance Board Member (1971)

14.14 Microprocessors are getting into everything. We won't be able to pick up a single piece of equipment in the near future, except maybe a broom, that hasn't got a microprocessor in it.

Arthur C. Clarke, American Author (1979)

14.15 People can still pull the computer's plug. However, we may have to work hard to even maintain that privilege.

Dr. Joseph Weizenbaum, American Computer Scientist/Sociologist (1979)

14.16 Even today I still get letters from young students here and there who say, Why are you people trying to program intelligence? Why don't you try to find a way to build a nervous system that will just spontaneously create it? Finally I decided that this was either a bad idea or else it would take thousands or millions of neurons to make it work and I couldn't afford to try to build a machine like that.

Marvin Minsky, American Computer Scientist (1982)

14.17 I think that intelligence does not emerge from a handful of very beautiful principles—like physics. It emerges from perhaps a hundred fundamentally different kinds of mechanisms that have to interact just right. So, even if it took only four years to understand them, it might take *four hundred* years to unscramble the whole thing.

Marvin Minsky, American Computer Scientist (1982)

14.18 People are just very complicated electronic mechanisms, and our emotions of love, hate, anger, and fear are wired into our brains. . . . I'm tinkering around inside the human computer.

Candace Pert, American Neuroscientist (1984)

15.
Cytology

Organisms are made up of cells much as
societies are made up of individual beings,
and for much the same reasons.

ISAAC ASIMOV

15.1 The whole organism subsists only by means of the reciprocal action of the single elementary parts.
Theodor Schwann, German Physiologist/Biologist (1810–1882)

15.2 All the varied forms in the animal tissues are nothing but transformed cells. . . . All my work has authorized me to apply to animals as to plants the doctrine of the individuality of the cells.
Theodor Schwann, German Physiologist/Biologist (1810–1882)

15.3 The cause of nutrition and growth resides not in the organism as a whole but in the separate elementary parts—the cells.
Theodor Schwann, German Physiologist/Biologist (1810–1882)

15.4 Every animal is a sum of vital units, each of which possesses the full characteristics of life. The character and the unity of life cannot be found in one definite point of a higher organization, for example, in the brain of man, but only in the definite, constantly recurring disposition shown individually by each single element. It follows that the composition of the major organism, the so-called individual, must be likened to a kind of social arrangement or society, in which a number of separate existences are dependent upon one another, in such a way, however, that each element possesses its own peculiar activity and carries out its own task by its own powers.
Rudolph Virchow, German Pathologist/Statesman (1821–1902)

15.5 Tissues evolve in time. A tissue consists of a society of complex organisms, which does not respond in an instantaneous manner to the changes of the environment. It may oppose such changes for a long time before adapting itself through slight or deep transformations. To study it at only one instant of its duration is almost meaningless. The temporal extension of a tissue is as important as its spatial existence.

Dr. Alexis Carroll, French-American Surgeon/Cytologist (1937)

15.6 The structure of tissues and their functions, are two aspects of the same thing.

Dr. Alexis Carroll, French-American Surgeon/Cytologist (1937)

15.7 The new cytology permits the identification of cells and the prediction of their conduct under given conditions. It reveals the specific properties of each type of cell. Thanks to this new cytology the mechanism of complex phenomena, which takes place in normal or pathological tissues, can be submitted to experimental analysis. Its fecundity will be necessarily greater than that of the classic cytology.

Dr. Alexis Carroll, French-American Surgeon/Cytologist (1937)

15.8 Structuro-functional wholeness or integrity, and specific structure, are actively built up and maintained in the course of development, chiefly by the morphogenetic and behavioral activity of cells or groups of cells.

E.S. Russell, English Scientist/Philosopher (1950)

15.9 I traveled among cells, watched their functioning . . . and realized that within myself was a grand assemblage of living organisms, all of which added up to me.

John Lilly, American Neurophysiologist/Spiritualist (1984)

15.10 Brain cells do selectively respond to features, but not uniquely so. The same cell responds to a color, a movement in a certain direction, the velocity of the movement, luminosity, and so on. Each cell is something like a person with many traits. So when you abstract blueness, you must address all the cells in the network that detect blue.

Karl Pribram, Austrian/American Neurophysiologist (1984)

15.11 Researchers now propose that oxidation processes are one cause of cellular aging. Put less prosaically, we may all be rusting.

Joanne Silberner, American Science Journalist (1986)

15.12 Needing help to spring to life, a virus is little more than a package of genetic information that must commandeer the machinery of a host cell to permit its own replication.

Peter Jaret, American Science Journalist (1986)

15.13 A miracle of evolution, the human immune system is not controlled by any central organ, such as the brain. Rather it has developed to function as a kind of biologic democracy, wherein the individual members achieve their ends through an information network of awesome scope.

Peter Jaret, American Science Journalist (1986)

16.

Dietetics

To many of us, the first law of dietetics seems to be: if it tastes good, it's bad for you.

ISAAC ASIMOV

16.1 Every man should eat and drink and enjoy the fruit of all his labor; it is the gift of God.

***The Bible* (circa 725 B.C.)**

16.2 Once it happened that all the other members of a man mutinied against the stomach, which they accused as the single, idle, uncontributing part in the entire body, while the rest were put to hardships and the expense of much labor to supply and minister to its appetites. However, the stomach merely ridiculed the fatuity of the members, who appeared not to be aware that the stomach certainly does receive the general nourishment, but only to return it again and distribute it amongst the rest.

Menenius Agrippa, Greek Philosopher (circa 500 B.C.)

16.3 The part of the soul which desires meats and drinks and the other things of which it has need by reason of the bodily nature, they (the gods) placed between the midriff and the boundary of the navel, contriving in all this region a sort of manager for the food of the body, and that there they bound it down like a wild animal which was chained up with man, and must be nourished if man is to exist.

Plato, Greek Philosopher (427 B.C.?–347 B.C.?)

16.4 Thou shouldst eat to live; not live to eat.

Marcus Tullius Cicero, Roman Orator/Statesman (106 B.C.–43 B.C.)

16.5 The chief pleasure [in eating] does not consist in costly seasoning, or exquisite flavor, but in yourself.

Horace (Quintus Horatius Flaccus), Roman Poet (65 B.C.–8 B.C.)

16.6 If you are surprised at the number of our maladies, count our cooks.

Lucius Annaeus Seneca (the Younger), Roman Philosopher/Statesman/Dramatist (4 B.C.?–65 A.D.)

16.7 Eating should be done in silence, lest the windpipe open before the gullet, and life be in danger.

***The Talmud* (circa 400)**

16.8 Well loved he garlic, onions, and eke leeks,
And for to drinken strong wine, red as blood.

Geoffrey Chaucer, English Poet (1340?–1400)

16.9 For its merit, I will knight it and make it sir-loin.
[Upon tasting the upper part of the loin.]

Charles II, English King (1630?–1685)

16.10 Bread is the staff of life.

Jonathan Swift, Irish Satirist/Clergyman (1667–1745)

16.11 A cucumber should be well sliced and dressed with pepper and vinegar, and then thrown out, as good for nothing.

Samuel Johnson, English Lexicographer/Poet/Critic (1709–1784)

16.12 A light supper, a good night's sleep, and a fine morning have often made a hero of the same man, who, by indigestion, a restless night, and a rainy morning would have proved a coward.

Edmund Burke, English Statesman/Author (1729–1797)

16.13 The discovery of a new dish does more for human happiness than the discovery of a new star.

Anthelme Brillat-Savarin, French Gastronomist (1755–1826)

16.14 Animals feed; man eats; the man of intellect alone knows how to eat.

Anthelme Brillat-Savarin, French Gastronomist (1755–1826)

16.15 The longer I live, the more I am convinced that the apothecary is of more importance than Seneca; and that half the unhappiness in the world proceeds from little stoppages, from a duct choked up, from food pressing in the wrong place, from a vext duodenum, or an agitated pylorus.

Sydney Smith, English Essayist/Clergyman (1771–1845)

16.16 The way to a man's heart is through his stomach.

Sara Payson Parton (Fanny Fern), English Author (1811–1872)

16.17 Beautiful soup! Who cares for fish
Game, or any other dish?
Who would not give all else for two
Pennyworth only of beautiful soup?
Lewis Carroll (Charles Lutwidge Dodgson), English Author/ Mathematician (1832–1898)

16.18 Hail, Gastronome, Apostle of Excess,
Well skilled to overeat without distress!
Thy great invention, the unfatal feast,
Shows Man's superiority to Beast.
Ambrose Bierce, American Satirist (1842–1914?)

16.19 We are all dietetic sinners; only a small percent of what we eat nourishes us, the balance goes to waste and loss of energy.
Sir William Osler, Canadian Physician/Anatomist (1849–1919)

16.20 He sows hurry and reaps indigestion.
Robert Louis Stevenson, Scottish Poet/Novelist (1850–1894)

16.21 Gave up spinach for Lent.
F. Scott Fitzgerald, American Novelist (1896–1940)

16.22 The human species was born when one isolated group of bipedal apes got itself stuck and then speciated to get better survival value out of eating meat.
Richard Leakey, English Paleontologist (1944–)

16.23 Food is the burning question in animal society, and the whole structure and activities of the community are dependent upon questions of food-supply.
Charles Elton, English Ecologist (1960)

16.24 The F.D.A. has so many rules that can be gotten around that the consumer has no protection at all. You never know what you're eating. I'm horrified when I discover the nature of ingredients in consumer products as a result of my scientific work.
Tina Chen, American Dietician (1971)

16.25 A moderate watcher of children's television programs receives over 5,000 messages per year, primarily advising him to eat snacks, sweets, and soda pop.
Robert B. Choate, American Nutritionist (1971)

16.26 My stomach is a laboratory.
Latzi Wittenberg, American Chef (1971)

16.27 Sure, cancer is a long and painful illness. But if you take the nitrites out of cured pork, you get botulism. That'll take you out fast and you'll have nothing to worry about.
Earl Butz, American Secretary of Agriculture (1975)

16.28 With the continued pressure on the Oglala to adopt mainstream ways, the ceremoniousness of the dog feast is likely to increase still more. The modern white man eats buffalo, deer, and elk, sometimes even horsemeat. But as long as the white man does not consume dog, he does not consume those who do.

William K. Powers and Marla N. Powers, American Anthropologist and American Sociologist (1986)

17.
The Earth

The earth is the only home that any of us have — so far, anyway.

ISAAC ASIMOV

17.1 For we are dwelling in a hollow of the earth, and fancy that we are on the surface. . . . But the fact is, that owing to our feebleness and sluggishness we are prevented from reaching the surface of the air.
Plato, Greek Philosopher (427 B.C.?–347 B.C.?)

17.2 It [the earth] alone remains immoveable, whilst all things revolve round it.
Pliny the Elder (Gaius Plinius Secundus), Roman Naturalist/Historian (23–79)

17.3 Since nothing stands in the way of the movability of the earth, I believe we must now investigate whether it also has several motions, so that it can be considered one of the planets. That it is not the center of all the revolutions is proven by the irregular motions of the planets, and their varying distances from the earth, which cannot be explained as concentric circles with the earth at the center.
Nicholas Copernicus, Polish Astronomer (1473–1543)

17.4 The Earth obey'd and straight
Op'ning her fertile womb, teem'd at a birth
Innumerous living creatures, perfect forms,
Limb'd and full grown.
John Milton, English Poet (1608–1674)

17.5 Within a finite period of time past, the earth must have been, and within a finite period of time to come the earth must again be, unfit for the habitation of man as at present constituted, unless operations have been, or are to be performed, which are impossible under the laws to which the known operations going on at present in the material world are subject.

Count Benjamin Thompson, English Physicist/Statesman (1753–1814)

17.6 [The great earthquake of 1835] at once destroyed our oldest idea that Earth is the very emblem of solidity, for it moved beneath our feet like a thin crust over a fluid. One second of time created in the mind a strange idea of insecurity, which hours of reflection would not have produced.

Charles Robert Darwin, English Naturalist/Evolutionist (1809–1882)

17.7 They say,
The solid earth whereon we tread

In tracts of fluent heat began,
And grew to seeming-random forms,
The seeming prey of cyclic storms,
Till at the last arose the Man. . . .

Alfred, Lord Tennyson, English Poet (1809–1892)

17.8 The earth in its rapid motion round the sun possesses a degree of living force so vast that, if turned into the equivalent of heat, its temperature would be rendered at least one thousand times greater than that of red-hot iron.

James Prescott Joule, English Physicist (1818–1889)

17.9 The present state of the earth and of the organisms now inhabiting it, is but the last stage of a long and uninterrupted series of changes which it has undergone, and consequently, that to endeavour to explain and account for its present condition without any reference to those changes (as has frequently been done) must lead to very imperfect and erroneous conclusions.

Alfred Russel Wallace, English Naturalist/Evolutionist (1823–1913)

17.10 Even if only one in a hundred of the ten billion suitable planets has actually got life well under way, there would be more than 100 million such planets. No, it is presumptuous to think that we are alone.

Harlow Shapley, American Astronomer (1885–1972)

17.11 We can hardly pick up a copy of a newspaper or magazine nowadays without being informed exactly how many million years ago some remarkable event in the history of the earth occurred.

Adolph Knopf, American Geologist (1941)

17.12 What beauty. I saw clouds and their light shadows on the distant dear earth. . . . The water looked like darkish, slightly gleaming spots. . . . When I watched the horizon, I saw the abrupt, contrasting transition from the earth's light-colored surface to the absolutely black sky. I enjoyed the rich color spectrum of the earth. It is surrounded by a light blue aureole that gradually darkens, becomes turquoise, dark blue, violet, and finally coal black. [Observations during the first manned orbit of the earth, April 12, 1961.]

Yuri Alekseyevich Gagarin, Russian Astronaut (1961)

17.13 When the earth came alive it began constructing its own membrane, for the general purpose of editing the sun.

Lewis Thomas, American Medical Educator/Writer (1974)

17.14 The key to the past lies in the record of the past, in the oral and written records, the record left in artefacts [sic], fossils, sediments, and so on. And in all those records, one consistent fact emerges. The Earth has been subjected to a series of catastrophes resulting from the fact that it has been tilted or turned completely upside-down. And that is the *only* possible explanation which fully accounts for the details in *all* of the records.

Peter Warlow, British Physicist/Author/Lecturer (1982)

17.15 According to the new theory [continental drift] about 300 million years ago, all the landmasses of the earth were joined on one supercontinent, which geologists refer to as Pangaea. About 200 million years ago, Pangaea began to break into two enormous landmasses, one in the south known as Gondwana, and the other, named Laurasia, in the north. The Gondwana landmass was made up of present-day Africa, Australia, India, and South America, with what is now Antarctica forming the centerpiece.

William J. Zinmeister, American Paleontologist/Geoscientist (1986)

17.16 The earth's rotation rate, which determines the length of day, can be sped up or slowed down by winds blowing on mountains. Similarly, "winds," or flow patterns of the earth's liquid outer core, can change the rotation rate by pushing on whatever bumps may be on the core-mantle boundary; the bigger the bump, the greater effect.

Stefi Weisburd, American Science Journalist (1987)

17.17 Studies of some accurately dated ancient and medieval eclipse records, principally those in which the observers made careful estimates of the time of day or night, reveal that the length of the day is gradually increasing. Although the rate of increase—about one-fiftieth of a second every thousand years—may seem trifling, in the million or so days that have elapsed since the earliest reliable astronomical records (700 B.C.), the earth has lost several hours compared with an ideal clock that would keep perfect time, neither gaining nor losing.

Richard F. Stephenson, American Physicist/Astronomer (1987)

18.
Ecology

What makes it so hard to organize the environment sensibly is that everything we touch is hooked up to everything else.

ISAAC ASIMOV

18.1 Nature, being inconstant and taking pleasure in creating and making constantly new lives and forms, because she knows that her terrestrial materials become thereby augmented, is more ready and more swift in her creating than time in his destruction; and so she has ordained that many animals shall be food for others. Nay, this not satisfying her desire, to the same end she frequently sends forth certain poisonous and pestilential vapors upon the vast increase and congregation of animals; and most of all upon men, who increase vastly because other animals do not feed upon them.
Leonardo da Vinci, Italian Architect/Artist/Inventor (1452–1519)

18.2 Land that is left wholly to nature . . . is called, as indeed it is, waste.
John Locke, English Philosopher (1632–1704)

18.3 Nature, who is a great economist, converts the recreation of one animal to the support of another.
Gilbert White, English Naturalist/Ecologist (1720–1793)

18.4 A few examples of local, and perhaps but one or two of absolute, extirpation of species can as yet be proved, and these only where the interference of man has been conspicuous.
Sir Charles Lyell, Scottish Geologist (1797–1875)

18.5 There is nothing in which the birds differ more from man than the way in which they can build and yet leave a landscape as it was before.
Robert Staughton Lynd, American Sociologist (1892–1970)

18.6 Unless man can make new and original adaptations to his environment as rapidly as his science can change the environment, our culture will perish.
Carl R. Rogers, American Psychologist (1902–1987)

18.7 The crescendo of noise—whether it comes from truck or jackhammer, siren or airplane—is more than an irritating nuisance. It intrudes on privacy, shatters serenity, and can inflict pain. We dare not be complacent about this ever mounting volume of noise. In the years ahead, it can bring even more discomfort—and worse—to the lives of people.
Lyndon Baines Johnson, American President (1908–1973)

18.8 Eventually, we'll realize that if we destroy the ecosystem, we destroy ourselves.
Jonas Salk, American Medical Researcher/Microbiologist (1915–)

18.9 They're too close to the trees to see the forest. People in California or New York understand that Alaska is not so big. They live in places where the wilderness once seemed limitless, but they know it disappears.
Edgar Wayburn, American Environmentalist (1970)

18.10 My people are not concerned about a phony economy like the Caucasians. We have our land and it has been good to us. The sea has been good to us. The Bureau of Indian Affairs has given us enough education to read about what has happened in the lower 48, what they have done to our land. Now they want to come up here and rape our land.
Joe Upicksoun, Native American Environmental Activist (1971)

18.11 Congress says on the one hand we must preserve our environment but on the other hand approves projects over which citizens express concern that they might damage the environment.
Barrington D. Parker, American Judge (1971)

18.12 The present fad is to talk about environment, but the clinician was concerned about environmental disease long before the word became something that almost everyone could pronounce.
Mitchell R. Zaron, American Physician/Educator (1971)

18.13 We [in New York] intend to redesign our entire purchasing system to include a general preference for recycled products.
John V. Lindsay, American Politician (1971)

18.14 The country is coming to understand the global dimensions of the environmental crisis. What is needed now is a precise pinpointing of the sources of environmental degradation, and tough-minded, long-term action to enhance the quality of our lives.

Joint Statement, American Environmental Group Leaders (1971)

18.15 At the turn of the century, reduction plants processed food wastes, dead animals, and fish for the production of glycerines, oils, fertilizers, and animal hides. Although each of the plants was successful initially, by 1915 all had been abandoned, not only for economic reasons, but also because of the uncontrollable nuisances they created.

George J. Kupchik, American Environmentalist/Administrator (1971)

18.16 A stable population with stable consumption patterns would still face increasing environmental problems.

Barry Commoner, American Biologist/Educator (1971)

18.17 Something like this ministry [Britain's ecology ministry] is going to have to happen in America. As your city situation deteriorates, new measures will be accepted.

Peter Edward Walker, British Ecology Minister (1971)

18.18 Once the Africans wore animal skins, and when they were discarded the ants disposed of them. Now the people here are starting to wear nylons, and the ants can no longer cope.

Bert de la Bat, South African Conservationist (1971)

18.19 This Is Our State. We Belong Here—Rockefeller Does Not—We Need Our Strip Mines.

Picket Sign, West Virginian Strip Miners and Supporters (1971)

18.20 Population size x Per capita consumption x Environmental impact per unit of production = Level of pollution.

Dr. Paul Ehrlich, American Biologist/Ecologist (1971)

18.21 It is not enough to treat chronic respiratory disease, and then send the victim back into the smog-filled air or the fumes and dust of an unwholesome work environment.

John J. Hanlon, American Assistant Surgeon General (1971)

18.22 We're going to let the public make some decisions about whether it's going to be sleek, smart, expensive automobiles or whether it's going to be alternative forms of transportation.

Jerome Kretchmer, American Environmental Protection Administrator (1971)

18.23 Three hundred trout are needed to support one man for a year. The trout, in turn, must consume 90,000 frogs, that must consume 27 million grasshoppers that live off of 1,000 tons of grass.

G. Tyler Miller, Jr., American Chemist (1971)

18.24 The highway system devours land resources and atmosphere at a rate that is impossible to sustain.

George W. Brown, American Business Executive (1972)

18.25 In its broadest ecological context, economic development is the development of more intensive ways of exploiting the natural environment.

Richard Wilkinson, American Economist/Historian (1973)

18.26 Industry has been compelled to spend more and more of its research dollars to comply with environmental, health, and safety regulations—and to move away from longer term efforts aimed at major scientific advances.

Rawleigh Warner, Jr., American Industrialist/Chairman, Mobil Oil (1973)

18.27 We prefer economic growth to clean air.

Charles Barden, American Environmental Administrator (1975)

18.28 I want to turn women loose on the environmental crisis. . . . Nobody knows more about pollution when detergents back up in the sink.

Nelson Rockefeller, American Politician (1975)

18.29 Wildlife management consists mainly of raising more animals for hunters to shoot.

George Schaller, American Zoologist/Ecologist/Educator (1984)

18.30 For what are the whales being killed? For a few hundred jobs and products that are not needed, since there are cheap substitutes. If this continues, it will be the end of living and the beginning of survival. The world is being totaled.

George Schaller, American Zoologist/Ecologist/Educator (1984)

18.31 Destroying species is like tearing pages out of an unread book, written in a language humans hardly know how to read, about the place where they live.

Rolston Holmes III, American Philosopher (1985)

18.32 This being the only living world we are ever likely to know, let us join to make the most of it.

Edward O. Wilson, American Entomologist/Sociobiologist (1985)

18.33 Few scientists were willing to jeopardize their research funds by publicly criticizing EPA's interpretation of the scientific record.

Roy Gould, American Ecologist (1986)

18.34 I do wonder what the devil we are doing killing 50,000 porpoises a year instead of developing a technology for tuna fishing that does not slaughter those animals. We are not very impressive as moralists. . . . Every now and then

I am reminded why the rest of the world won't listen to us when it comes to conservation: We don't lead the world, we just talk as if we did.
Roger Caras, American Newspaper Columnist (1986)

18.35 Parasites are four times as effective as predators in controlling pests.
Peter Spinks, American Science Journalist (1986)

19.

Electricity and Electronics

The difference between electricity

and electronics is the difference between a

toaster and a television set.

ISAAC ASIMOV

19.1 Do not electric bodies by friction emit a subtile exhalation or spirit by which they perform their attractions?
Isaac Newton, English Physicist/Mathematician (1642–1727)

19.2 A turkey is to be killed for our dinner by the *electrical shock* and roasted by the *electrical jack,* before a fire kindled by the *electrified bottle:* when the healths of all the famous electricians in *England, Holland, France,* and *Germany* are to be drank in *electrified bumpers,* under the discharge of guns from the *electrical battery.*
Benjamin Franklin, American Inventor/Statesman (1706–1790)

19.3 I am busy just now again on electro-magnetism, and think I have got hold of a good thing, but can't say. It may be a weed instead of a fish that, after all my labor, I may at last pull up.
Michael Faraday, English Chemist/Physicist (1791–1867)

19.4 It is a fact—or have I dreamt it—that by means of electricity the world of matter has become a great nerve, vibrating thousands of miles in a breathless point of time?
Nathaniel Hawthorne, American Author (1804–1864)

19.5 The farthest Thunder that I heard
Was nearer than the Sky
And rumbles still, though torrid Noons
Have lain their missiles by—
The Lightning that preceded it
Struck no one but myself—
But I would not exchange the Bolt
For all the rest of Life—
Indebtedness to Cxygen
The Happy may repay,
But not the obligation
To Electricity—
It founds the Homes and decks the Days
And every clamor bright
Is but the gleam concomitant
Of that waylaying Light—
The Thought is quiet as a Flake—
A Crash without a Sound,
How Life's reverberation
Is Explanation found—
Emily Dickinson, American Poet (1830–1886)

19.6 The experimental investigation by which Ampere established the law of the mechanical action between electric currents is one of the most brilliant achievements in science. The whole theory and experiment, seems as if it had leaped, full grown and full armed, from the brain of the "Newton of Electricity." It is perfect in form, and unassailable in accuracy, and it is summed up in a formula from which all the phenomena may be deduced, and which must always remain the cardinal formula of electro-dynamics.
James Clerk Maxwell, Scottish Physicist/Mathematician (1831–1879)

19.7 The electric light invades the dunnest deep of Hades.
Cries Pluto, 'twixt his snores: "O tempora! O mores!"
Ambrose Bierce, American Satirist (1842–1914?)

19.8 There is no plea which will justify the use of high-tension and alternating currents, either in a scientific or a commercial sense. They are employed solely to reduce investment in copper wire and real estate. . . . I have always consistently opposed high-tension and alternating systems of electric lighting, not only on account of danger, but because of their general unreliability and unsuitability for any general system of distribution.
Thomas Alva Edison, American Inventor (1847–1931)

19.9 We have a system [alternating current] whereby the deadly electricity of the alternating current can do no harm unless a man is fool enough to swallow a whole dynamo.
George Westinghouse, American Mechanical Engineer (1884)

19.10 Just as certain as death, Westinghouse will kill a customer within six months after he puts in a [alternating current] system of any size.

Thomas Alva Edison, American Inventor (1847–1931)

19.11 It was necessary to invent everything. Dynamos, regulators, meters, switches, fuses, fixtures, underground conductors with their necessary connecting boxes, and a host of other detail parts, even down to insulating tape. [On his invention of an electric distribution system for electric light.]

Thomas Alva Edison, American Inventor (1847–1931)

19.12 In fact, I've come to the conclusion that I never did know anything about it [electricity].

Thomas Alva Edison, American Inventor (1847–1931)

19.13 The first World System power plant can be put into operation in nine months. With this power plant it will be practical to attain electrical activities up to ten million horsepower and it is designed to serve for as many technical achievements as are possible without undue expense. Among these may be mentioned:

1. Interconnection of the existing telegraph exchanges of offices all over the world;
2. Establishment of a secret and noninterferable government telegraph service;
3. Interconnection of all the present telephone exchanges or offices all over the Globe;
4. Universal distribution of general news, by telegraph or telephone, in connection with the Press;
5. Establishment of a World System of intelligence transmission for exclusive private use;
6. Interconnection and operation of stock tickers of the world;
7. Establishment of a World System of musical distribution, etc.;
8. Universal registration of time by cheap clocks indicating the time with astronomical precision and requiring no attention whatever;
9. Facsimile transmission of typed or handwritten characters, letters, checks, etc.;
10. Establishment of a universal marine service enabling navigation of all ships to steer perfectly without compass, to determine the exact location, hour, and speed, to prevent collisions and disasters, etc.;
11. Inauguration of a system of world printing on land and sea;
12. Reproduction anywhere in the world of photographic pictures and all kinds of drawings or records.

Nikola Tesla, American Electrical Engineer/Inventor (1856–1943)

19.14 The momentous laws of induction between currents and between currents and magnets were dicovered by Michael Faraday in 1831–32. Faraday was asked: "What is the use of this discovery?" He answered: "What is the use of a child—it grows to be a man." Faraday's child has grown to be a man and is now the basis of all the modern applications of electricity.

Alfred North Whitehead, English Mathematician/Philosopher (1861–1947)

19.15 They could have done it better with an ax.
[On the ineffective results of the first use of the electric chair, the execution of William Kemmler, condemned murderer.]

George Westinghouse, American Mechanical Engineer (1890)

19.16 In the early days of telephone engineering, the mere sending of a message was so much of a miracle that nobody asked how it should be sent.

Norbert Wiener, American Inventor/Mathematician (1894–1964)

19.17 Modern electronic communications and information processing are marvelous extensions of man's senses and mind. But these same technologies are producing closer, more complex interactions between different peoples, and between people and machines, without the integrating force of common social purpose.

Jack Morton, American Engineer (1913–1971)

19.18 There is nothing wrong with electricity; nothing except that modern man is not a god who holds the thunderbolts but a savage who is struck by lightning.

Gilbert Keith Chesterton, English Essayist/Novelist/Poet (1930)

19.19 Between 1880 and 1920, the first and second generations of men who created and ran the modern electrical industry formed the vanguard of science-based industrial development in the United States. These were the people who first successfully combined the discoveries of physical science with the mechanical know-how of the workshop to produce the much heralded electrical revolution in power generation, lighting, transportation, and communication: who forged the great companies which manufactured that revolution and the countless electric utilities, electric railways, and telephone companies which carried it across the nation.

David Noble, American Engineer/Educator (1977)

19.20 Sometime between 1740 and 1780, electricians were for the first time enabled to take the foundations for their field for granted. From that point they pushed on to more concrete and recondite problems.

Thomas Kuhn, American Engineer/Educator (1977)

19.21 Just as the Industrial Revolution enabled man to apply and control greater physical power than his own muscle could provide, so electronics has extended his intellectual power.

Dr. Robert Noyce, American Electronic Engineer (1977)

20.
Embryology

The fertilized ovum of a mouse and a whale look much alike, but differences quickly show up in the course of their development. If we could study their molecules with the naked eyes, we would see the differences from the start.

ISAAC ASIMOV

20.1 This work must begin with the conception of man, and describe the nature of the womb and how the fetus lives in it, up to what stage it resides there, and in what way it quickens into life and feeds. Also its growth and what interval there is between one stage of growth and another. What it is that forces it out from the body of the mother, and for what reasons it sometimes comes out of the mother's womb before the due time.

Leonardo da Vinci, Italian Architect/Artist/Inventor (1452–1519)

20.2 Philosophers have been of the opinion that our immortal part acquires during this life certain habits of action or of sentiment, which become forever indissoluble, continuing after death in future state of existence . . . I would apply this ingenious idea to the generation, or production of the embryo, or new animal, which partakes so much of the form and propensities of the parent.

Erasmus Darwin, English Physician/Poet (1731–1802)

20.3 Man has two conditions of existence in the body. Hardly two creatures can be less alike than an infant and a man. The whole fetal state is a preparation for birth. . . . The human brain, in its earlier stage, resembles that of a fish: as it is developed, it resembles more the cerebral mass of a reptile; in its increase, it is like that of a bird, and slowly, and only after birth, does it assume the proper form and consistence of the human encephalon.

Charles Bell, Scottish Anatomist/Surgeon (1774–1842)

20.4 The embryos of mammals, of birds, lizards, and snakes are, in their earliest states, exceedingly like one another, both as a whole and in the mode of development of their parts, indeed we can often distinguish such embryos only by their size. I have two little embryos in spirit [alcohol] to which I have omitted to attach the names. I am now quite unable to say to what class they belong.

Karl Ernst von Baer, German Biologist/Embryologist (1792–1876)

20.5 Embryology furnishes the best measure of the true affinities between animals.

Jean Louis Agassiz, Swiss/American Naturalist/Geologist (1807–1873)

20.6 Embryology will often reveal to us the structure, in some degree obscured, of the prototype of each great class.

Charles Robert Darwin, English Naturalist/Evolutionist (1809–1882)

20.7 Ontogeny recapitulates phylogeny.

Ernst Heinrich Haeckel, German Biologist/Philosopher (1834–1919)

20.8 Here we have a baby. It is composed of a bald head and a pair of lungs.

Eugene Field, American Poet/Humorist (1850–1895)

20.9 As the human fetus develops, its changing form seems to retrace the whole of human evolution from the time we were cosmic dust to the time we were single-celled organisms in the primordial sea to the time we were four-legged, land-dwelling reptiles and beyond, to our current status as large-brained, bipedal mammals. Thus, humans seem to be the sum total of experience since the beginning of the cosmos.

Jonas Salk, American Medical Researcher/Microbiologist (1915–)

20.10 Even in the earliest stages of embryogeny, if differentiation has not proceeded too far, regulatory activities will lead to the production of a normal embryo from a half or a quarter of the segmenting egg.

E.S. Russell, English Scientist/Philosopher (1950)

21. Energy

The law of conservation of energy tells us we can't get something for nothing, but we refuse to believe it.

ISAAC ASIMOV

21.1 Heat is a motion; expansive, restrained, and acting in its strife upon the smaller particles of bodies. But the expansion is thus modified; while it expands all ways, it has at the same time an inclination upward. And the struggle in the particles is modified also; it is not sluggish, but hurried and with violence.

Francis Bacon, English Philosopher/Essayist/Statesman (1561–1626)

21.2 The total energy of the universe is constant and the total entropy is continually increasing.
[The two laws of thermodynamics.]

Isaac Newton, English Physicist/Mathematician (1642–1727)

21.3 About ten years ago I read . . . that the celebrated Amontons, using a thermometer of his own invention, had discovered that water boils at a fixed degree of heat. I was at once inflamed with a great desire to make for myself a thermometer of the same sort, so that I might with my own eyes perceive this beautiful phenomenon of nature, be convinced of the truth of the experiment. . . . I gathered that a thermometer might be perhaps constructed with mercury, which would not be so hard to construct, and by the use of which it might be possible to carry out the experiment which I so greatly desired to try. When a thermometer of that sort was made (perhaps imperfect in many ways) the result answered to my prayer; and with great pleasure of mind I observed the truth of the thing.

Gabriel Daniel Fahrenheit, German Physicist (1686–1736)

21.4 As the ostensible effect of the heat . . . consists not in warming the surrounding bodies but in rendering the ice fluid, so, in the case of boiling, the heat absorbed does not warm surrounding bodies but converts the water into vapor. In both cases, considered as the cause of warmth, we do not perceive its presence: it is concealed, or latent, and I gave it the name of "latent heat."

Joseph Black, Scottish Physician/Physicist/Chemist (1728–1799)

21.5 Heat cannot be lessened or absorbed without the production of living force, or its equivalent attraction through space.

James Prescott Joule, English Physicist (1818–1889)

21.6 The heat produced in maximal muscular effort, continued for twenty minutes, would be so great that, if it were not promptly dissipated, it would cause some of the albuminous substances of the body to become stiff, like a hard-boiled egg.

Walter B. Cannon, American Physiologist (1871–1945)

21.7 Whenever there is a great deal of energy in one region and very little in a neighboring region, energy tends to travel from the one region to the other, until equality is established. This whole process may be described as a tendency towards democracy.

Bertrand Russell, English Philosopher/Mathematician (1872–1970)

21.8 $E = mc^2$

[E = Energy, m = mass, c = velocity of light.]

Albert Einstein, German/American Physicist (1879–1955)

21.9 A theory is more impressive the greater is the simplicity of its premises, the more different are the kinds of things it relates, and the more extended its range of applicability. Therefore, the deep impression which classical thermodynamics made on me. It is the only physical theory of universal content which I am convinced, that within the framework of applicability of its basic concepts will never be overthrown.

Albert Einstein, German/American Physicist (1879–1955)

21.10 Entropy is time's arrow.

Sir Arthur Stanley Eddington, English Astronomer/Mathematician (1882–1944)

21.11 An atomic power plant is somewhat analogous to a steam boiler—it could blow up.

You want to put a safety valve on it and you want a good safety valve. However, what has happened was that, even though excellent safety devices had been proposed and were installed, the question was raised eventually, "Is there any remaining risk?" Now you can always hypothesize that some little boy sits on the valve, or that sabotage occurs, or just assume it doesn't work. . . . The fellow who runs the reactor—he doesn't want to blow himself up. He is going to put in the best safety devices, and the idea of assuming that all these fail is being ridiculously cautious.

Kenneth S. Pitzer, American Director of Atomic Energy Commission (1952)

21.12 Entropy is a measure of the heat energy in a substance that has been lost and is no longer available for work. It is a measure of the deterioration of a system.

***Popular Science Encyclopedia of the Sciences* (1963)**

21.13 Suppose we take a quantity of heat and change it into work. In doing so, we haven't destroyed the heat, we have only transferred it to another place or perhaps changed it into another energy form.

Isaac Asimov, American Biochemist/Author (1965)

21.14 Conversion to exclusive dependence on solar energy would clearly require major changes in our technology and economy in the direction of greater frugality and decentralization.

William Ophuls, American Ecologist (1977)

21.15 Solar energy can generate some net, concentrated energy in the form of food, fiber, and electricity, but the amount per area is small because most of the solar energy is consumed by the various structures that have to be maintained and operated to collect and concentrate the energy.

Howard Odum, American Environmentalist (1978)

21.16 In our very distant future we are likely to find that there is only *one* energy, which has manifold expressions, depending on the state of consciousness that interacts with the energy. However, we presently have a scientific foundation which has already segmented and delineated uniquely different energy characteristics as perceived by our biological senses, and by our extended instrumentation senses. Thus we must continue along the path already laid down by our scientific forebearers, until we have reached the level of consciousness where the unity can be known.

William Tiller, American Science Educator (1981)

22. Engineering

Science can amuse and fascinate us all,

but it is engineering that changes the world.

ISAAC ASIMOV

22.1 [Engineers are] the direct and necessary instrument of coalition by which alone the new social order can commence.

Auguste Comte, French Philosopher (1798–1857)

22.2 Thousands of engineers can design bridges, calculate strains and stresses, and draw up specifications for machines, but the great engineer is the man who can tell whether the bridge or the machine should be built at all, where it should be built, and when.

Eugene G. Grace, American Industrialist (1876–1960)

22.3 Scientists study the world as it is, engineers create the world that never has been.

Theodore von Karman, Hungarian/American Aeronautical Engineer (1881–1963)

22.4 To define it rudely, but not inaptly, engineering is the art of doing that well with one dollar which any bungler can do with two after a fashion.

Arthur M. Wellington, American Economist (1887)

22.5 The theory of control in engineering, whether human or animal or mechanical, is a chapter in the theory of messages.

Norbert Weiner, American Inventor/Mathematician (1894–1964)

22.6 In the conception of a machine or a product of a machine, there is a point where one may leave off for parsimonious reasons, without having reached aesthetic perfection; at this point perhaps every mechanical factor is accounted for, and the sense of incompleteness is due to the failure to recognize the claims of the human agent. Aesthetics carries with it the implications of alternatives between a number of mechanical solutions of equal validity; and unless this awareness is present at every stage of the process . . . it is not likely to come out with any success in the final stage of design.

Lewis Mumford, American Social Philosopher/Writer (1895–)

22.7 One has to watch out for engineers—they begin with sewing machines and end up with the atomic bomb.

Marcel Pagnol, French Playwright/Film Critic (1895–1974)

22.8 On one occasion committee members were asked by the chairman, who was also in charge of the project, to agree that a certain machine be run at a power which was ten percent lower than the design value. [Franz Eugen] Simon objected, arguing that "design value" should mean what it said. Thereupon the chairman remarked, "Professor Simon, don't you see that we are not talking about science, but about engineering, which is an art." Simon was persistent: "What would happen if the machine were run at full power?" "It might get too hot." "But, Mr. Chairman," came Simon's rejoinder, "Can't artists use thermometers?"

Nicholas Kurti, British Physicist (1908–)

22.9 The ultimate object of design is form.

Christopher Alexander, American Engineer (1964)

22.10 A scientist can discover a new star but he cannot make one. He would have to ask an engineer to do it for him.

Gordon L. Glegg, American Engineer (1969)

22.11 As never before, the work of the engineer is basic to the kind of society to which our best efforts are committed. Whether it be city planning, improved health care in modern facilities, safer and more efficient transportation, new techniques of communication, or better ways to control pollution and dispose of wastes, the role of the engineer—his initiative, creative ability, and hard work—is at the root of social progress.

Richard Milhous Nixon, American President (1971)

22.12 Nature provides her creatures with the best possible systems; the engineer has to try to implement them in devices.

Robert Zinter, American Optical Engineer (1986)

23.
Entomology

Human beings can easily destroy every elephant on earth, but we are helpless against the mosquito.

ISAAC ASIMOV

23.1 Go to the ant, thou sluggard; consider her ways, and be wise.
The Bible (circa 325 A.D.)

23.2 The spring before [1632], espetially at the month of 1 May, ther was such a quantitie of a great sorte of flies, like (for bignes) to wasps, or bumble-bees, which came out of holes in the ground, and replenished all the woods, and eate the green-things, and made such a constante yelling noyes, as made all the woods ring of them, and ready to deafe the hearers. They have not by the English been heard or seen before or since. But the Indeans told them that sicknes would follow, and so it did in June, July, August, and the cheefe heat of sommer.
William Bradford, American Governor (1590–1657)

23.3 How doth the little busy bee
Improve each shining hour,
And gather honey all the day
From every opening flower.
Isaac Watts, English Hymn Writer/Theologian (1674–1748)

23.4 Philosophers more grave than wise
Hunt science down in butterflies;
Or fondly poring on a spider,
Stretch human contemplation wider.
John Gay, English Poet/Dramatist (1685–1732)

23.5 In the summer, our meadows and fields are beautifully illuminated by an immense number of fire-flies, which in the calm of the evening sweetly wander here and there at a small distance from the ground. By their alternate glows of light, they disseminate a kind of universal splendour, which, being always contrasted with the darkness of the night, has a most surprising effect. I have often read by their assistance; that is, I have taken one carefully by the wings and, carrying it along the lines of my book, I have, when thus assisted by these living flambeaux, perused whole pages and then thankfully dismissed these little insect-stars.

Hector St. John de Crevecoeur, American Farmer/French Consul to New York (1735–1813)

23.6 Ye ugly, creepin', blastit wonner,
Detested, shunn'd by saunt an' sinner!
How dare ye set your fit upon her,
Sae fine a lady?
Gae somewhere else, and seek your dinner
On some poor body.

[To a louse.]

Robert Burns, Scottish Poet (1759–1796)

23.7 Even bees, the little alms-men of spring bowers,
Know there is richest juice in poison-flowers.

John Keats, English Poet (1795–1821)

23.8 One day, on tearing off some old bark [from a tree], I saw two rare beetles, and seized one in each hand. Then I saw a third and new kind, which I could not bear to lose, so I popped the one which I held in my right hand into my mouth. Alas! it ejected some intensely acrid fluid, which burnt my tongue so that I was forced to spit the beetle out, which was lost, as was the third one.

Charles Robert Darwin, English Naturalist/Evolutionist (1809–1882)

23.9 A noiseless, patient spider,
I mark'd, where on a little promontory, it stood, isolated;
Mark'd how, to explore the vacant, vast surrounding,
It launch'd forth filament, filament, filament, out of itself
Ever unreeling them—ever tirelessly speeding them.

Walt Whitman, American Poet (1819–1892)

23.10 None but a naturalist can understand the intense excitement I experienced when at last I captured it [a hitherto unknown species of butterfly]. My heart began to beat violently, the blood rushed to my head, and I felt much more like fainting than I have done when in apprehension of immediate death. I had a headache the rest of the day, so great was the excitement produced by what will appear to most people a very inadequate cause.

Alfred Russel Wallace, English Naturalist/Evolutionist (1823–1913)

23.11 For among Bees and Ants are social systems found
so complex and well-order'd as to invite offhand
a pleasant fable enough: that once upon a time,
or ever a man were born to rob their honeypots,
bees were fully endow'd with Reason and only lost it
by ordering so their life as to dispense with it;
whereby it pined away and perish'd of disuse.

Robert Bridges, English Poet (1844–1930)

23.12 If Darwin were alive today the insect world would delight and astound him with its impressive verification of his theories of the survival of the fittest. Under the stress of intensive chemical spraying the weaker members of the insect populations are being weeded out. . . . Only the strong and fit remain to defy our efforts to control them.

Rachel Carson, American Marine Biologist/Author (1907–1964)

23.13 The most intensely social animals can only adapt to group behavior. Bees and ants have no option when isolated, except to die. There is really no such creature as a single individual; he has no more life of his own than a cast-off cell marooned from the surface of your skin.

Lewis Thomas, American Medical Educator/Writer (1974)

23.14 305 species of insects, mites, and ticks [are] known to possess genetic strains resistant to one or more chemical pesticides.

U.S. Council on Environmental Quality, Ninth Annual Report (1978)

23.15 The acceleration of a flea is twenty times more powerful than that of a moon rocket reentering the earth's atmosphere.

Margaret Rothschild, English Entomologist/Naturalist (1984)

23.16 Harvester and carpenter ants . . . will send out lone foragers that recruit their nestmates only after having found food. But army ants . . . have many strategies for extremely rapid foraging. Able to communicate messages chemically among thousands of ants, they can mobilize a sizeable and replenishable strike force at a target within seconds.

But if the offensive capabilities of these predator ants are remarkable, the defenses of their prey are equally impressive. Our most recent studies reveal that target ant colonies resort to civil defense tactics, chemical warfare, counterattack, and even a version of the "MX missile ploy" (with ant colonies periodically moving to one of a series of new nesting quarters).

Howard Topoff, American Psychologist (1987)

24.
Error

A subtle thought that is in error may yet give rise to fruitful inquiry that can establish truths of great value.

ISAAC ASIMOV

24.1 The cautious seldom err.
Confucius, Chinese Philosopher (551 B.C.–479 B.C.)

24.2 To stumble twice against the same stone is a proverbial disgrace.
Marcus Tullius Cicero, Roman Orator/Statesman (106 B.C.–43 B.C.)

24.3 Truth comes out of error more readily than out of confusion.
Francis Bacon, English Philosopher/Essayist/Statesman (1561–1626)

24.4 Errors like straws upon the surface flow:
He who would search for pearls must dive below.
John Dryden, English Poet/Critic/Dramatist (1631–1700)

24.5 To err is human, to forgive divine.
Alexander Pope, English Poet/Satirist (1688–1744)

24.6 When every one is in the wrong, every one is in the right.
Pierre Auguste Claude Nivelle de la Chaussée, French Dramatist (1692–1754)

24.7 It is error only, and not truth, that shrinks from inquiry.
Thomas Paine, English Author/Political Theorist (1737–1809)

24.8 A man's errors are what make him amiable.
Johann Wolfgang von Goethe, German Poet/Dramatist/Novelist (1749–1832)

24.9 Much as I venerate the name of Newton, I am not therefore obliged to believe that he was infallible. I see . . . with regret that he was liable to err, and that his authority has, perhaps, sometimes even retarded the progress of science.

Thomas Young, English Physician/Writer (1773–1829)

24.10 Errors using inadequate data are much less than those using no data at all.

Charles Babbage, English Inventor/Mathematician (1792–1871)

24.11 Any pride I might have felt in my conclusions was perceptibly lessened by the fact that I knew that the solution of these problems had almost always come to me as the gradual generalization of favorable examples, by a series of fortunate conjectures, after many errors.

Hermann Ludwig Ferdinand von Helmholtz, German Physicist/ Physiologist (1821–1894)

24.12 The man who makes no mistakes does not usually make anything.

Edward John Phelps, American Diplomat/Lawyer (1822–1900)

24.13 Irrationally held truths may be more harmful than reasoned errors.

Thomas Henry Huxley, English Biologist/Evolutionist (1825–1895)

24.14 Next to the promulgation of the truth, the best thing I can conceive that man can do is the public recantation of an error.

Joseph Lister, English Surgeon (1827–1912)

24.15 What is called science today consists of a haphazard heap of information, united by nothing, often utterly unnecessary, and not only failing to present one unquestionable truth, but as often as not containing the grossest errors, today put forward as truths, and tomorrow overthrown.

Leo N. Tolstoy, Russian Author (1828–1910)

24.16 Give me a fruitful error any time, full of seeds, bursting with its own corrections. You can keep your sterile truth for yourself.

Vilfredo Pareto, Italian Economist/Sociologist (1848–1923)

24.17 There are sadistic scientists who hurry to hunt down error instead of establishing the truth.

Marie Curie, Polish Chemist/Physicist (1867–1934)

24.18 Looking back . . . over the long and labyrinthine path which finally led to the discovery [of the quantum theory], I am vividly reminded of Goethe's saying that men will always be making mistakes as long as they are striving after something.

Albert Einstein, German/American Physicist (1879–1955)

24.19 In the field of thinking, the whole history of science from geocentrism to the Copernican revolution, from the false absolutes of Aristotle's physics to the relativity of Galileo's principle of inertia and to Einstein's theory of relativity, shows that it has taken centuries to liberate us from the systematic errors, from the illusions caused by the immediate point of view as opposed to "decentered" systematic thinking.

Jean Piaget, Swiss Psychologist (1896–1980)

25.
Evolution

Those who reject biological evolution do so, usually, not out of reason, but out of unjustified vanity.

ISAAC ASIMOV

25.1 Men first appeared as fish. When they were able to help themselves they took to land.
Anaximander, Greek Philosopher (611 B.C.–547 B.C.)

25.2 Many races of living creatures must have been unable to continue their breed: for in the case of every species that now exists, either craft, or courage, or speed, has from the beginning of its existence protected and preserved it.
Empedocles, Greek Philosopher (490 B.C.?–430 B.C.)

25.3 How like to us is that filthy beast the ape.
Marcus Tullius Cicero, Roman Orator/Statesman (106 B.C.–43 B.C.)

25.4 Observe constantly that all things take place by change, and accustom thyself to consider that the nature of the universe loves nothing so much as to change the things which are, and to make new things like them.
Marcus Aurelius, Roman Emperor/Philosopher (121–180)

25.5 I died as mineral and became a plant.
I died as plant and rose to animal.
I died as animal and I became man.
Why should I fear? When was I less by dying?
Jalaoddin Rumi, Persian Mystic/Theologian (1207–1273)

25.6 It is suitable to the magnificent harmony of the universe that the species of creatures should, by gentle degrees, ascend upward from us toward His perfection, as we see them gradually descend from us downward.

John Locke, English Philosopher (1632–1704)

25.7 Vast chain of being! which from God began,
Natures ethereal, human, angel, man,
Beast, bird, fish, insect, what no eye can see,
No glass can reach; from Infinite to thee,
From thee to nothing.—On superior pow'rs
Were we to press, inferior might on ours;
Or in the full creation leave a void,
Where, one step broken, the great scale's destroy'd;
From Nature's chain whatever link you strike,
Tenth, or ten thousandth, breaks the chain alike.

Alexander Pope, English Poet/Satirist (1688–1744)

25.8 Accustom infancy to the inclemency of the weather and the vicissitude of the seasons, inured to fatigue and compelled idleness, and without arms to defend their lives and prey against other ferocious animals, or to escape from them by speed of foot, men become robust and so vigorous as scarcely to be affected by external circumstances. The children of such parents bring into the world their excellent constitution, and strengthening it by the same exercises, acquire in this way all the vigor of which the human species is capable.

Jean-Jacques Rousseau, Swiss/French Philosopher/Author (1712–1778)

25.9 The operations of Nature in the production of animals show that there is a primary and predominant cause which gives to animal life the power of progressive organization, of gradually complicating and perfecting not only the organism as a whole, but each system of organs in particular.

Jean-Baptiste de Monet, Chevalier de Lamarck, French Naturalist (1744–1829)

25.10 We know that this animal, the tallest of mammals, dwells in the interior of Africa, in places where the soil, almost always arid and without herbage, obliges it to browse on trees and to strain itself continuously to reach them. This habit sustained for long, has had the result in all members of its race that the forelegs have grown longer than the hind legs and that its neck has become so stretched, that the giraffe, without standing on its hind legs, lifts its head to a height of six meters.

Jean-Baptiste de Monet, Chevalier de Lamarck, French Naturalist (1744–1829)

25.11 [Darwinism] is a system which is so repugnant at once to history, to the tradition of all peoples, to exact science, to observed facts, and even to Reason herself, [it] would seem to need no refutation, did not alienation from

God and the leaning toward materialism, due to depravity, eagerly seek a support in all this tissue of fables.

Pope Pius IX (1792–1878)

25.12 But the idea that any of the lower animals have been concerned in any way with the origin of man—is not this degrading? Degrading is a term, expressive of a notion of the human mind, and the human mind is liable to prejudices which prevent its notions from being invariably correct. Were we acquainted for the first time with the circumstances attending the production of an individual of our race, we might equally think them degrading, and be eager to deny them, and exclude them from the admitted truths of nature.

Robert Chambers, Scottish Publisher/Editor (1802–1871)

25.13 My father was a creole, his father a Negro, and his father a monkey; my family, it seems, begins where yours left off.

Alexandre Dumas (père), French Novelist/Dramatist (1802–1870)

25.14 According to my derivative hypothesis, a change takes place first in the structure of the animal, and this, when sufficiently advanced, may lead to modifications of habits. . . . "Derivation" holds that every species changes, in time, by virtue of inherent tendencies thereto. "Natural Selection" holds that no such change can take place without the influence of altered external circumstances educing or selecting such change. . . . The hypothesis of "natural selection" totters on the extension of a conjectural condition, explanatory of extinction to the majority of organisms, and not known or observed to apply to the origin of any species.

Richard Owen, English Paleontologist/Biologist (1804–1892)

25.15 Mankind's struggle upwards, in which millions are trampled to death, that thousands may mount on their bodies.

Clara Lucas Balfour, English Social Reformer (1808–1878)

25.16 Man is descended from a hairy, tailed quadruped, probably arboreal in its habits.

Charles Robert Darwin, English Naturalist/Evolutionist (1809–1882)

25.17 We must, however, acknowledge, as it seems to me, that man with all his noble qualities . . . still bears in his bodily frame the indelible stamp of his lowly origin.

Charles Robert Darwin, English Naturalist/Evolutionist (1809–1882)

25.18 The expression often used by Mr. Herbert Spencer of "the Survival of the Fittest" is more accurate, and is sometimes equally convenient.

Charles Robert Darwin, English Naturalist/Evolutionist (1809–1882)

25.19 Thus, from the war of nature, from famine and death, the most exalted object which we are capable of conceiving, namely, the production of the higher animals, directly follows. There is grandeur in this view of life, with its several powers, having been originally breathed by the Creator into a few forms or into one; and that, whilst this planet has gone cycling on according to the fixed law of gravity, from so simple a beginning endless forms most beautiful and most wonderful have been, and are being evolved.

Charles Robert Darwin, English Naturalist/Evolutionist (1809–1882)

25.20 The main conclusion arrived at in this work, namely, that man is descended from some lowly organized form, will, I regret to think, be highly distasteful to many. But there can hardly be a doubt that we are descended from barbarians.

Charles Robert Darwin, English Naturalist/Evolutionist (1809–1882)

25.21 If my body come from brutes, though somewhat finer than their own,
I am heir, and this my kingdom. Shall the royal voice be mute?
No, but if the rebel subject seek to drag me from the throne,
Hold the scepter, Human Soul, and rule the province of the brute.

Alfred, Lord Tennyson, English Poet (1809–1892)

25.22 In due time the evolution theory will have to abate its vehemence, cannot be allow'd to dominate everything else, and will have to take its place as a segment of the circle, the cluster—as but one of many theories, many thoughts, of profoundest value—and readjusting the differentiating much, yet leaving the divine secrets just as inexplicable and unreachable as before—maybe more so.

Walt Whitman, American Poet (1819–1892)

25.23 Some filosifers think that a fakkilty's granted
The minnit it's felt to be thoroughly wanted.

That the fears of a monkey whose holt chanced to fail
Drawed the vertibry out to a prehensile tail.

James Russell Lowell, American Poet/Critic/Diplomat (1819–1891)

25.24 This survival of the fittest, which I have here sought to express in mechanical terms, is that which Mr. Darwin has called "natural selection, or the preservation of favored races in the struggle for life."

Herbert Spencer, English Philosopher/Psychologist (1820–1903)

25.25 From the remotest past which Science can fathom, up to the novelties of yesterday, that in which Progress essentially consists, is the transformation of the homogeneous into the heterogeneous.

Herbert Spencer, English Philosopher/Psycholgist (1820–1903)

25.26 The conditions that direct the order of . . . the living world . . . are marked by their persistence in improving the birthright of successive generations. They determine, at much cost of individual comfort, that each plant and animal shall, on the general average, be endowed at its birth with more suitable natural faculties than those of its representative in the preceding generation.

Francis Galton, English Geneticist/Anthropologist (1822–1911)

25.27 The problem [evolution] presented itself to me, and something led me to think of the positive checks described by Malthus in his Essay on Population, a work I had read several years before, and which had made a deep and permanent impression on my mind. These checks—war, disease, famine, and the like—must, it occurred to me, act on animals as well as man. Then I thought of the enormously rapid multiplication of animals, causing these checks to be much more effective in them than in the case of man; and while pondering vaguely on this fact, there suddenly flashed upon me the idea of the survival of the fittest—that the individuals removed by these checks must be on the whole inferior to those that survived. I sketched the draft of my paper . . . and sent it by the next post to Mr. Darwin.

Alfred Russel Wallace, English Naturalist/Evolutionist (1823–1913)

25.28 It is an error to imagine that evolution signifies a constant tendency to increased perfection. That process undoubtedly involves a constant remodeling of the organism in adaption to new conditions; but it depends on the nature of those conditions whether the direction of the modifications effected shall be upward or downward.

Thomas Henry Huxley, English Biologist/Evolutionist (1825–1895)

25.29 If, then, the question is put to me, would I rather have a miserable ape for a grandfather, or a man highly endowed by nature and possessing great means and influence, and yet who employs those faculties and that influence for the mere purpose of introducing ridicule into a grave scientific discussion—I unhesitatingly affirm my preference for the ape.

Thomas Henry Huxley, English Biologist/Evolutionist (1825–1895)

25.30 Every individual alive today, the highest as well as the lowest, is derived in an unbroken line from the first and lowest forms.

August Frederick Leopold Weismann, German Biologist/Geneticist (1834–1914)

25.31 The theory of evolution has often been perverted so as to indicate that what is merely animal and brutal must gain the ascendancy. The contrary seems to me to be the case, for in man it is the spirit, and not the body, which is the deciding factor.

August Frederick Leopold Weismann, German Biologist/Geneticist (1834–1914)

25.32 Nothing is constant but change! All existence is a perpetual flux of "being and becoming!" That is the broad lesson of the evolution of the world.

Ernst Heinrich Haeckel, German Biologist/Philosopher (1834–1919)

25.33 When you were a tadpole and
I was a fish in Palaeozoic time
And side by side in the sluggish tide,
we sprawled in the ooze and slime.
Langdon Smith, American Journalist (1858–1908)

25.34 What shall we say of the intelligence, not to say religion, of those who are so particular to distinguish between fishes and reptiles and birds, but put a man with an immortal soul in the same circle with the wolf, the hyena, and the skunk? What must be the impression made upon children by such a degradation of man?
William Jennings Bryan, American Politician/Lawyer (1860–1925)

25.35 There is no more reason to believe that man descended from some inferior animal than there is to believe that a stately mansion has descended from a small cottage.
William Jennings Bryan, American Politician/Lawyer (1860–1925)

25.36 Children, behold the Chimpanzee;
He sits in the ancestral tree
From which we sprang in ages gone.
I'm glad we sprang: had we held on,
We might, for aught that I can say,
Be horrid Chimpanzees today.
Oliver Herford, American Poet/Humorist (1863–1935)

25.37 The dodo never had a chance. He seems to have been invented for the sole purpose of becoming extinct and that was all he was good for.
William Cuppy, American Critic/Humorist (1884-1949)

25.38 The majority of evolutive movements are degenerative. Progressive cases are exceptional. Characters appear suddenly that have no meaning in the atavistic series. Evolution in no way shows a general tendency toward progress. . . . The only thing that could be accomplished by slow changes would be the accumulation of neutral characteristics without value for survival. Only important and sudden mutations can furnish the material which can be utilized by selection.
John Burdon Sanderson Haldane, English Geneticist (1892–1964)

25.39 If today you can take a thing like evolution and make it a crime to teach it in the public schools, tomorrow you can make it a crime to teach it in the private schools, and next year you can make it a crime to teach it to the hustings or in the church. At the next session you may ban books and the newspapers. . . . Ignorance and fanaticism are ever busy and need feeding. Always feeding and gloating for more. Today it is the public school teachers; tomorrow the private. The next day the preachers and the lecturers, the magazines, the books, the newspapers. After a while, Your Honor, it is the setting of man against man and creed against creed until with flying banners and beating drums we are marching backward to the glorious ages of the sixteenth century when bigots lighted fagots to burn the men who dared to bring any intelligence and enlightenment and culture to the human mind. [At the Scopes Trial, Dayton, Ohio.]

Clarence Darrow, American Lawyer (1925)

25.40 The net effect of Clarence Darrow's great speech yesterday seems to be precisely the same as if he had bawled it up a rainspout in the interior of Afghanistan.

H.L. Mencken, American Journalist/Critic (1925)

25.41 It shall be unlawful for any teacher in any of the universities, normals, and all other public schools in the state, which are supported in whole or in part by the public school funds of the state, to teach the theory that denies the story of the divine creation of man as taught in the Bible, and to teach instead that man has descended from a lower order of animals.

Act of the Legislature of Tennessee (March 21, 1925)

25.42 If we are correct in understanding how evolution actually works, and provided we can survive the complications of war, environmental degradation, and possible contact with interstellar planetary travelers, we will look exactly the same as we do now. We won't change at all. The species is now so widely dispersed that it is not going to evolve, except by gradualism.

Richard Leakey, English Paleontologist (1944–)

25.43 The teaching of the Church leaves the doctrine of evolution an open question.

Pope Pius XII (1950)

25.44 Today, the theory of evolution is an accepted fact for everyone but a fundamentalist minority, whose objections are based not on reasoning but on doctrinaire adherance [sic] to religious principles.

James D. Watson, American Geneticist/Biophysicist (1965)

25.45 If we sink to the biochemical level, then the human being has lost a great many synthetic abilities possessed by other species and, in particular, by plants and microorganisms. Our loss of ability to manufacture a variety of vitamins makes us dependent on our diet and, therefore, on the greater synthetic versatility of other creatures. This is as much a "degenerative" change as the tapeworm's abandonment of a stomach it no longer needs, but since we are prejudiced in our own favor, we don't mention it.

Isaac Asimov, American Biochemist/Author (1976)

25.46 This monkey mythology of Darwin is the cause of permissiveness, promiscuity, pills, prophylactics, perversions, abortions, pornography, pollution, poisoning, and proliferation of crimes of all types.

Judge Braswell Dean, American Judge, Atlanta, Georgia (1981)

25.47 I believe that life can go on forever. It takes a million years to evolve a new species, ten million for a new genus, one hundred million for a class, a billion for a phylum—and that's usually as far as your imagination goes. In a billion years, it seems, intelligent life might be as different from humans as humans are from insects. But what would happen in another ten billion years? It's utterly impossible to conceive of ourselves changing as drastically as that, over and over again. All you can say is, on that kind of time scale the material form that life would take is completely open. To change from a human being to a cloud may seem a big order, but it's the kind of change you'd expect over billions of years.

Freeman Dyson, American Mathematician (1986)

26. Experiment

Experimentation is the least arrogant method of gaining knowledge. The experimenter humbly asks a question of nature.

ISAAC ASIMOV

26.1 No one tests the depth of a river with both feet.
African Proverb (Date unknown)

26.2 In dealing with a scientific problem, I first arrange several experiments, since my purpose is to determine the problem in accordance with experience, and then to show why the bodies are compelled so to act. That is the method which must be followed in all researches upon the phenomena of Nature.
Leonardo da Vinci, Italian Architect/Artist/Inventor (1452–1519)

26.3 Mr. Hobbes told me that the cause of his Lordship's [Francis Bacon's] death was trying an experiment: viz., as he was taking the air in a coach with Dr. Witherborne, a Scotchman, physician to the King, towards Highgate, snow lay on the ground, and it came into my Lord's thoughts, why flesh might not be preserved in snow as in salt. They were resolved they would try the experiment presently. They alighted out of the coach and went into a poor woman's house at the bottom of Highgate Hill and bought a hen and made the woman exenterate it, and then stuffed the body with snow, and my Lord did help to do it himself. The snow so chilled him that he immediately fell so extremely ill that he could not return to his lodgings.
John Aubrey, English Author/Biographer (1626-1697)

26.4 It is the weight, not numbers of experiments that is to be regarded.
Isaac Newton, English Physicist/Mathematician (1642–1727)

26.5 There is no experiment to which a man will not resort to avoid the real labor of thinking.
Joshua Reynolds, English Portrait Painter (1723–1792)

26.6 The observer listens to nature: the experimenter questions and forces her to reveal herself.
Georges Cuvier, French Zoologist/Anatomist (1769–1832)

26.7 The true worth of an experimenter consists in his pursuing not only what he seeks in his experiment, but also what he did not seek.
Claude Bernard, French Physiologist (1813–1878)

26.8 We must never make experiments to confirm our ideas, but simply to control them.
Claude Bernard, French Physiologist (1813–1878)

26.9 The first experiment a child makes is a physical experiment: the suction-pump is but an imitation of the first act of every new-born infant.
John Tyndall, English Physicist (1820–1893)

26.10 It is the care we bestow on apparently trifling, unattractive detail and very troublesome minutiae which determines the result.
Theobald Smith, American Pathologist (1859–1934)

26.11 The mere formulation of a problem is far more often essential than its solution, which may be merely a matter of mathematical or experimental skill. To raise new questions, new possibilities, to regard old problems from a new angle requires creative imagination and marks real advances in science.
Albert Einstein, German/American Physicist (1879–1955)

26.12 The experiment serves two purposes, often independent one from the other: it allows the observation of new facts, hitherto either unsuspected, or not yet well defined; and it determines whether a working hypothesis fits the world of observable facts.
René J. Dubos, French/American Bacteriologist (1901–1982)

26.13 To test a perfect theory with imperfect instruments did not impress the Greek philosophers as a valid way to gain knowledge.
Isaac Asimov, American Biochemist/Author (1965)

26.14 In the behavioral sciences, when people are working with live human subjects, the biases and theoretical preconceptions of the experimenter cannot help but influence, and often profoundly, the behavior of the subject being studied.
Leon Kamin, American Psychologist/Educator (1986)

26.15 There is a kind of arrogance in ascribing naivete to the great minds of previous generations. Newton and his successors were well aware of friction, turbulence, and many other subtleties which affected the exactness of measurements.

Mark and Maxine Bridger, American Mathematicians/Science Journalists (1986)

27.
Fact

It is the easiest thing in the world to

deny a fact. People do it all the time.

Yet it remains a fact just the same.

ISAAC ASIMOV

27.1 Science is the knowledge of consequences, and dependence of one fact upon another.

Thomas Hobbes, English Philosopher (1588–1679)

27.2 Facts are stubborn things.

Tobias George Smollett, English Novelist (1721–1771)

27.3 Facts are to the mind the same thing as food to the body. On the due digestion of facts depends the strength and wisdom of the one, just as vigor and health depend on the other. The wisest in council, the ablest in debate, and the most agreeable in the commerce of life is that man who has assimilated to his understanding the greatest number of facts.

Edmund Burke, English Statesman/Author (1729–1797)

27.4 If I set out to prove something, I am no real scientist—I have to learn to follow where the facts lead me—I have to learn to whip my prejudices.

Lazzaro Spallanzani, Italian Physiologist/Microbiologist (1729–1799)

27.5 Facts are things which cannot be altered or disputed.

Robert Burns, Scottish Poet (1759–1796)

27.6 I grow daily to honor facts more and more, and theory less and less.

Thomas Carlyle, Scottish Historian/Philosopher (1795–1881)

27.7 No facts are to me sacred; none are profane; I simply experiment, an endless seeker, with no past at my back.

Ralph Waldo Emerson, American Essayist/Philosopher/Poet (1803–1882)

27.8 I must begin with a good body of facts and not from a principle (in which I always suspect some fallacy) and then as much deduction as you please.

Charles Robert Darwin, English Naturalist/Evolutionist (1809–1882)

27.9 Negative facts when considered alone never teach us anything.

Claude Bernard, French Physiologist (1813–1878)

27.10 Blessed is the man who, having nothing to say, abstains from giving in words evidence of the fact.

George Eliot (Mary Ann Evans), English Novelist (1819–1880)

27.11 The brightest flashes in the world of thought are incomplete until they have been proved to have their counterparts in the world of fact.

John Tyndall, English Physicist (1820–1893)

27.12 Every fact that is learned becomes a key to other facts.

Edward Livingston Youmans, American Chemist (1821–1887)

27.13 Precise facts alone are worthy of science.

Jean Henri Fabre, French Entomologist/Naturalist (1823–1915)

27.14 I am afraid that there is something empirical in the French mind and that the only way to make it admit a truth is to present this truth as an experimental fact.

Jules Lachelier, French Philosopher (1832–1918)

27.15 There is something fascinating about science. One gets such wholesome returns of conjectures out of such trifling investment of fact.

Mark Twain (Samuel Langhorne Clemens), American Journalist/ Novelist (1835–1910)

27.16 Get your facts first, then you can distort them as much as you please.

Mark Twain (Samuel Langhorne Clemens), American Journalist/ Novelist (1835–1910)

27.17 If all single facts, all separate phenomena, were as directly accessible to us as we demand that knowledge of them to be, science would never have arisen.

Ernst Mach, Austrian Physicist (1838–1916)

27.18 The growth [of science] consists in a continual analysis of facts, of rough and general observation into groups of facts more precise and minute.

Walter Pater, English Critic/Essayist (1839–1894)

27.19 "Facts, facts, facts," cries the scientist if he wants to emphasize the necessity of a firm foundation for science. What is a fact? A fact is a thought that is true. But the scientist will surely not recognize something which depends on men's varying states of mind to be the firm foundation of science.
Gottlob Frege, German Mathematician/Logician (1848–1925)

27.20 Fed on the dry husks of facts, the human heart has a hidden want which science cannot supply.
Sir William Osler, Canadian Physician/Anatomist (1849–1919)

27.21 Facts are the air of science. Without them a man of science can never rise. Without them your theories are vain surmises. But while you are studying, observing, experimenting, do not remain content with the surface of things. Do not become a mere recorder of facts, but try to penetrate the mystery of their origin. Seek obstinately for the laws that govern them.
Ivan Petrovich Pavlov, Russian Physiologist (1849–1936)

27.22 Science is facts. Just as houses are made of stones, so is science made of facts. But a pile of stones is not a house and a collection of facts is not necessarily science.
Jules Henri Poincaré, French Mathematician/Astronomer (1854–1912)

27.23 The aim of science is to seek the simplest explanations of complex facts. We are apt to fall into the error of thinking that the facts are simple because simplicity is the goal of our quest. The guiding motto in the life of every natural philosopher should be, "Seek simplicity and distrust it."
Alfred North Whitehead, English Mathematician/Philosopher (1861–1947)

27.24 Science robs men of wisdom and usually converts them into phantom beings loaded up with facts.
Miguel de Unamuno, Spanish Philosopher/Author (1864–1936)

27.25 I pass with relief from the tossing sea of Cause and Theory to the firm ground of Result and Fact.
Sir Winston Churchill, English Statesman/Author (1874–1965)

27.26 We are Marxists, and Marxism teaches that in our approach to a problem we should start from objective facts, not from abstract definitions, and that we should derive our guiding principles, policies, and measures from an analysis of these facts.
Mao Tse Tung, Chinese Political Leader (1893–1976)

27.27 I'm not afraid of facts, I welcome facts, *but a congeries of facts is not equivalent to an idea*. This is the essential fallacy of the so-called "scientific" mind. People who mistake facts for ideas are incomplete thinkers; they are gossips.
Cynthia Ozick, American Writer (1931-)

27.28 Facts are the world's data. Theories are structures of ideas that explain and interpret facts.
Stephen Jay Gould, American Biologist/Author (1983)

27.29 Here are the opinions on which my facts are based.
Anonymous (1986)

28.
Forestry

Humanity is cutting down its forests,

apparently oblivious to the fact that

we may not be able to live without them.

ISAAC ASIMOV

28.1 Even the gods dwelt in the woods.
Virgil (Publius Vergilius Maro), Roman Poet (70 B.C.–19 B.C.)

28.2 To linger silently among the healthful woods, musing on such things as are worthy of a wise and good man.
Horace (Quintus Horatius Flaccus), Roman Poet (65 B.C.–8 B.C.)

28.3 The monarch oak, the patriarch of the trees,
Shoots rising up, and spreads by slow degrees:
Three centuries he grows, and three he stays
Supreme in state; and in three more decays.
John Dryden, English Poet/Critic/Dramatist (1631–1700)

28.4 Trees perspire profusely, condense largely, and check evaporation so much that woods are always moist: no wonder, therefore, that they contribute much to pools and streams.
Gilbert White, English Naturalist/Ecologist (1720–1793)

28.5 One impulse from a vernal wood
May teach you more of man,
Of moral evil and of good,
Than all the sages can.
William Wordsworth, English Poet (1770–1850)

28.6 This is the forest primeval.
Henry Wadsworth Longfellow, American Poet (1807–1882)

28.7 Men have been talking now for a week at the post office about the age of the great elm, as a matter interesting but impossible to be determined. The very choppers and travelers have stood upon its prostrate trunk and speculated upon its age, as if it were a profound mystery. I stooped and read its years to them (127 at nine and a half feet), but they heard me as the wind that once sighed through its branches. They still surmised that it might be two hundred years old, but they never stooped to read the inscription. Truly they love darkness rather than light. One said it was probably one hundred and fifty, for he had heard somebody say that for fifty years the elm grew, for fifty it stood still, and for fifty it was dying. (Wonder what portion of his career he stood still!) Truly all men are not men of science. They dwell within an integument of prejudice thicker than the bark of the cork-tree, but it is valuable chiefly to stop bottles with. Tied to their buoyant prejudices, they keep themselves afloat when honest swimmers sink.
Henry David Thoreau, American Writer/Naturalist (1817–1862)

28.8 If [in a rain forest] the traveler notices a particular species and wishes to find more like it, he must often turn his eyes in vain in every direction. Trees of varied forms, dimensions, and colors are around him, but he rarely sees any of them repeated. Time after time he goes towards a tree which looks like the one he seeks, but a closer examination proves it to be distinct.
Alfred Russel Wallace, English Naturalist/Evolutionist (1823–1913)

28.9 The forests of America, however slighted by man, must have been a great delight to God; for they were the best he ever planted.
John Muir, American Naturalist/Conservationist (1838–1914)

28.10 There is in us all the dark conscience of our murder of the primeval forests—of something in their depth which is a depth in us; of our refusal of their value, of our disdain of the red man who was the spirit of those forests and who is yet, beneath the layers of law and memory, the spirit of ourselves.
Waldo Frank, American Novelist/Social Critic (1889–1967)

28.11 I think that I shall never see
A billboard lovely as a tree
Indeed, unless the billboards fall
I'll never see a tree at all.
Ogden Nash, American Humorist/Poet (1902–1971)

28.12 Every farm woodland, in addition to yielding lumber, fuel, and posts, should provide its owner a liberal education. This crop of wisdom never fails, but it is not always harvested.
Aldo Leopold, German Ecologist (1949)

28.13 Like the trees [in a rain forest] the lianas are also struggling to get upward. Too weak of themselves, they need the shoulders of others. Ever on the lookout for some object to grip, they curl or twine themselves around their victims, or shoot forth roots that cling into the bark, or hook themselves to objects by curved spines, or stretch out armlike projections that grip the sides of stems and trunks. In the end by some contrivance or another they get up and free themselves into the light.

R.W. G. Hingston, English Naturalist (1950)

28.14 Many parts of Northwestern Europe had achieved a kind of saturation with humankind by the 14th century. The great frontier boom that began about 900 led to a replication of manors and fields across the face of the land until, at least in the most densely inhabited regions, scant forests remained. Since woodlands were vital for fuel and as a source of building materials, mounting shortages created severe problems for human occupancy.

William McNeill, American Historian (1976)

28.15 It is estimated that thirty-five to fifty acres of rain forest are chopped down every minute.

George Schaller, American Zoologist/Ecologist/Educator (1984)

28.16 Pine is the larynx of the wind. No other trees unravel, comb, and disperse moving air so thoroughly. Yet they also seem to concentrate the winds, wringing mosaics of sound from gale weather—voice echoes, cries, sobs, conversations, maniacal calls. With the help of only slight imagination, they are receiving stations to which all winds check in, filtering out loads of B flats and F minors, processing auditory debris swept from all corners of the sound-bearing world.

John Eastman, American Author (1986)

29. Genetics

Genetics seems to be everything to those who have convinced themselves they have arisen from worthy ancestors.

ISAAC ASIMOV

29.1 If we had a thorough knowledge of all the parts of the seed of any animal (e.g., man), we could from that alone, by reasons entirely mathematical and certain, deduce the whole conformation and figure of each of its members, and, conversely, if we knew several peculiarities of this conformation, we would from those deduce the nature of its seed.

René Descartes, French Mathematician/Philosopher (1596–1650)

29.2 The ovary of an ancestress will contain not only her daughter, but also her granddaughter, her great-grand-daughter, and her great-great-grand-daughter, and if it is once proved that an ovary can contain many generations, there is no absurdity in saying that it contains them all.

Albrecht von Haller, Swiss Embryologist/Physician (1708–1777)

29.3 The fertilized germ of one of the higher animals . . . is perhaps the most wonderful object in nature. . . . On the doctrine of reversion [atavism] . . . the germ becomes a far more marvelous object, for, besides the visible changes which it undergoes, we must believe that it is crowded with invisible characters . . . separated by hundreds or even thousands of generations from the present time: and these characters, like those written on paper with invisible ink, lie ready to be evolved whenever the organization is disturbed by certain known or unknown conditions.

Charles Robert Darwin, English Naturalist/Evolutionist (1809–1882)

29.4 Breed is stronger than pasture.

George Eliot (Mary Ann Evans), English Novelist (1819–1880)

29.5 The theory is confirmed that pea hybrids form egg and pollen cells, which, in their constitution, represent in equal numbers all constant forms which result for the combination of the characters united in fertilization.

Gregor Johann Mendel, Austrian Botanist/Abbot (1822–1884)

29.6 Attributes of organisms consist of distinct, separate, and independent units.

Hugo Marie De Vries, Dutch Botanist (1848–1935)

29.7 Within the nucleus [of a cell] is a network of fibers, a sap fills the interstices of the network. The network resolves itself into a definite number of threads at each division of the cell. These threads we call chromosomes. Each species of animals and plants possesses a characteristic number of these threads which have definite size and sometimes a specific shape and even characteristic granules at different levels. Beyond this point our strongest microscopes fail to penetrate.

Thomas Hunt Morgan, American Geneticist (1866–1945)

29.8 The function of mutation is to maintain the stock of genetic variance at a high level.

Ronald Aylmer Fisher, English Geneticist (1890–1962)

29.9 Mutations merely furnish random raw material for evolution, and rarely, if ever determine the course of the process.

Sewall Wright, American Biologist/Geneticist (1892–1964)

29.10 THE DNA MOLECULE
THE DNA MOLECULE
THE DNA MOLECULE
is The Nude Descending a Staircase
a circular one.
See the undersurfaces
of the spiral
treads and
the spaces
in between.

May Swenson, American Poet (1919–)

29.11 The way to eliminate the unfit is to keep them from being born. . . . We should not only check degeneration—negatively—but further evolution, positively, by artificial insemination and work for the production of a nobler and nobler race of beings.

Hermann Joseph Muller, American Geneticist (1920)

29.12 Within the last five or six years [from 1916], from a common wild species of fly, the fruit fly, *Drosophila ampelophila,* which we have brought into the laboratory, have arisen over a hundred and twenty-five new types whose origin is completely known.

Thomas Hunt Morgan, American Geneticist (circa 1922)

29.13 Race involves the inheritance of similar physical variations by large groups of mankind, but its psychological and cultural connotations, if they exist, have not been ascertained by science.

American Anthropological Association, Resolution (1938)

29.14 No very radical changes in classification of the great branches of mankind are suggested when they are compared by the gene method.

L.C. Dunn, American Geneticist (1951)

29.15 "Random mating" obviously does not mean promiscuity; it simply means . . . that in the choice of mates for marriage there is neither preference for nor aversion to the union of persons similar or dissimilar *with respect to a given trait or gene*. Not all gentlemen prefer blondes or brunettes.

Bruce Wallace and Theodosius Dobzhansky, American Geneticist and Russian/American Anthropologist (1959)

29.16 Scientists are going to discover many subtle genetic factors in the makeup of human beings. Those discoveries will challenge the basic concepts of equality on which our society is based. Once we can say that there are differences between people that are easily demonstrable at the genetic level, then society will have to come to grips with understanding diversity—and we are not prepared for that.

David Baltimore, American Microbiologist (1983)

29.17 I love female intelligences. Every single cell in your body has two X chromosomes. Every cell in my body has one X chromosome and a crippled X chromosome, an X chromosome with an arm missing, called a Y chromosome. You women are so well balanced with your two X's. You can be grounded and do the gardening and take care of the kids and give them nurture, but we males have got to go out and explore the universe, banging our heads together and shooting one another.

John Lilly, American Neurophysiologist/Spiritualist (1984)

29.18 In a purely technical sense, each species of higher organism is richer in information than a Caravaggio painting, Bach fugue, or any other great work of art. Consider the typical case of the house mouse, *Mus musculus.* Each of its cells contains four strings of DNA, each of which comprises about a billion nucleotide pairs organized into a hundred thousand structural nucleotide pairs, organized into a hundred thousand structural genes. . . . The full information therein, if translated into ordinary-sized printed letters, would just about fill all 15 editions of the *Encyclopaedia Britannica* published since 1768.

Edward O. Wilson, American Entomologist/Sociobiologist (1985)

29.19 Only time and money stand between us and knowing the composition of every gene in the human genome.

Francis Crick, British Biophysicist/Geneticist (1986)

29.20 The dichotomies of nature versus nurture, constitutional versus acquired, and heredity versus environment reflect a bipolarity that does not exist. The bipolarity is a false one, since both social and biological influences affect behavior through the central nervous system, irrespective of how they gain their entry—either internally, by way of genetics, for instance, or externally, by way of stimuli transmitted through the senses from the environment.

Anke A. Ehrhardt, American Psychiatrist/Educator (1986)

30.
Geology

The earth is a book in which we read not only its history, but the history of the living things it has borne.

ISAAC ASIMOV

30.1 One frequently sees on high mountains conches and oyster shells, sometimes embedded in rocks. These rocks in pristine times were earth, and the shell fish and oysters lived in water. Subsequently everything was inverted. Things from the bottom came to the top, and the soft became hard. Careful considerations of these facts will lead to far-reaching conclusions.
Chu Hsi, Chinese Philosopher (1130–1200)

30.2 The whole terrestrial globe was taken to pieces at the Flood, and the strata now visible settled down from the promiscuous mass as any earthy sediment from a fluid.
John Woodward, English Physician/Geologist (1665–1728)

30.3 What clearer evidence could we have had of the different formation of these rocks, and of the long interval which separated their formation, had we actually seen them emerging from the bosom of the deep? . . . The mind seemed to grow giddy by looking so far into the abyss of time.
James Hutton, Scottish Geologist (1726–1797)

30.4 It was one thing to declare that we had not yet discovered the trace of a beginning, and another to deny that the earth ever had a beginning.
John Playfair, Scottish Mathematician/Geologist (1748–1819)

30.5 He plucks the pearls that stud the deep
Admiring Beauty's lap to fill;

He breaks the stubborn Marble's sleep,
Rocks disappear before his skill:
With thoughts that swell his glowing soul
He bids the ore illume the page,
And, proudly scorning Time's control,
Commences with an unborn age.

Charles Sprague, American Poet/Essayist/Banker (1791–1875)

30.6 The gradual advance of Geology, during the last twenty years to the dignity of a science, has arisen from the laborious and extensive collection of facts, and from the enlightened spirit in which the inductions founded on those facts have been deduced and discussed. To those who are unacquainted with this science, or indeed to any person not deeply versed in the history of this and kindred subjects, it is impossible to convey a just impression of the nature of that evidence by which a multitude of conclusions are supported:—evidence in many cases so irresistible, that the records of the past ages, to which it refers, are traced in language more imperishable that that of the historian of any human transactions; the relics of those beings, entombed in the strata which myriads of centuries have heaped upon their graves, giving a present evidence of their past existence, with which no human testimony can compete.

Charles Babbage, English Inventor/Mathematician (1792–1871)

30.7 It may undoubtedly be said that strata have been always forming somewhere, and therefore at every moment of past time Nature has added a page to her archives.

Sir Charles Lyell, Scottish Geologist (1797–1875)

30.8 We are as much gainers by finding a new property in the old earth as by acquiring a new planet.

Ralph Waldo Emerson, American Essayist/Philosopher/Poet (1803–1882)

30.9 One naturally asks, what was the use of this great engine set at work ages ago to grind, furrow, and knead over, as it were, the surface of the earth? We have our answer in the fertile soil which spreads over the temperate regions of the globe. The glacier was God's great plough.

Jean Louis Agassiz, Swiss/American Naturalist/Geologist (1807–1873)

30.10 We may confidently come to the conclusion that the forces that slowly and by little start the uplift of continents, and that . . . pour forth volcanic matter from open vents, are identical.

Charles Robert Darwin, English Naturalist/Evolutionist (1809–1882)

30.11 Daily it is forced home on the mind of the geologist that nothing, not even the wind that blows, is so unstable as the level of the crust of this Earth.

Charles Robert Darwin, English Naturalist/Evolutionist (1809–1882)

30.12 Geology gives us a key to the patience of God.

Josiah Gilbert Holland, American Novelist/Poet (1819–1881)

30.13 The facts proved by geology are briefly these: that during an immense, but unknown period, the surface of the earth has undergone successive changes; land has sunk beneath the ocean, while fresh land has risen up from it; mountain chains have been elevated; islands have been formed into continents, and continents submerged till they have become islands; and these changes have taken place, not once merely, but perhaps hundreds, perhaps thousands of times.

Alfred Russel Wallace, English Naturalist/Evolutionist (1823–1913)

30.14 When the curtain was then first raised that had veiled the history of the earth, and men, looking beyond the brief span within which they had supposed that history to have been transacted, beheld the records of a long vista of ages stretching far away into a dim illimitable past, the prospect vividly impressed their imagination. Astronomy had made known the immeasurable fields of space; the new science of geology seemed now to reveal boundless distances of time.

Sir Archibald Geikie, Scottish Geologist (1835–1924)

30.15 Reality is never skin-deep. The true nature of the earth and its full wealth of hidden treasures cannot be argued from the visible rocks, the rocks upon which we live and out of which we make our living. The face of the earth, with its upstanding continents and depressed ocean-deeps, its vast ornament of plateau and mountain-chain, is molded by structure and process in hidden depths.

Reginald A. Daly, American Geologist/Geophysicist (1871–1957)

30.16 The interpretation of messages from the earth's interior demands all the resources of ordinary physics and of extraordinary mathematics. The geophysicist is of a noble company, all of whom are reading messages from the untouchable reality of things. The inwardness of things—atoms, crystals, mountains, planets, stars, nebulas, universes—is the quarry of these hunters of genius and Promethean boldness.

Reginald A. Daly, American Geologist/Geophysicist (1871–1957)

30.17 The discovery of the famous original [Rosetta Stone] enabled Napoleon's experts to begin the reading of Egypt's ancient literature. In like manner the seismologists, using the difficult but manageable Greek of modern physics, are beginning the task of making earthquakes tell the nature of the earth's interior and translating into significant speech the hieroglyphics written by the seismograph.

Reginald A. Daly, American Geologist/Geophysicist (1871–1957)

30.18 The birth of a volcanic island is an event marked by prolonged and violent travail; the forces of the earth striving to create, and all the forces of the sea opposing.

Rachel Carson, American Marine Biologist/Author (1961)

30.19 In attempting to determine the real age of the earth, it should always be remembered, of course, that recorded history began only several thousand years ago. Not even uranium dating is capable of experimental verification, since no one could actually watch uranium decaying for millions of years to see what happens.

H.M. Morris, American Scientific Creationist/Author (1974)

30.20 Geologists have long since abandoned the idea that the Earth formed as a ball of liquid rock. We now know for sure that the Earth accumulated from solids. Partial melting caused a relatively thin crust of lighter rocks to form, segregated from the largely solid mantle that forms the outer half of the radius of the Earth, or seven-eighths of our planet's volume. In this picture, one can no longer rule out the possibility that volatile fluids could be "cooked out" from this mantle and work their way up towards the surface.

Tom Gold, American Radiophysicist/Educator (1986)

31. Gravity

We all know we fall. Newton's discovery was that the moon falls, too—and by the same rule that we do.

ISAAC ASIMOV

31.1 [The Elements] are mutually bound together, the lighter being restrained by the heavier, so that they cannot fly off; while, on the contrary, from the lighter tending upwards, the heavier are so suspended, that they cannot fall down. Thus, by an equal tendency in an opposite direction, each of them remains in its appropriate place, bound together by the never-ceasing revolution of the world.

Pliny the Elder (Gaius Plinius Secundus), Roman Naturalist/Historian (23–79)

31.2 I think gravity is nothing else than a natural appetency, given to the parts by the Divine Providence of the Maker of the universe, in order that they may establish their unity and wholeness by combining in the form of a sphere. It is probable that this affection also belongs to the sun, moon, and the planets, in order that they may, by its efficacy, remain in their roundness.

Nicholas Copernicus, Polish Astronomer (1473–1543)

31.3 I have not been able to discover the cause of those properties of gravity from phenomena, and I frame no hypotheses . . . it is enough that gravity does really exist, and act according to the laws which we have explained, and abundantly serves to account for the motions of the celestial bodies.

Isaac Newton, English Physicist/Mathematician (1642–1727)

31.4 That one body may act upon another at a distance through a vacuum without the mediation of anything else, by and through which their action and force may be conveyed from one to another, is to me so great an absurdity that, I believe, no man who has in philosophic matters a competent faculty of thinking could ever fall into it.

Isaac Newton, English Physicist/Mathematician (1642–1727)

31.5 Is it not true that the doctrine of attraction and gravity has done nothing but astonish our imagination?

Frederick the Great, King of Prussia (1712–1786)

31.6 When Newton saw an apple fall, he found . . .
A mode of proving that the earth turn'd round
In a most natural whirl, called "gravitation";
And thus is the sole mortal who could grapple,
Since Adam, with a fall or with an apple.

Lord Byron (George Gordon), English Poet/Dramatist (1788–1824)

31.7 Simple as the law of gravity now appears, and beautifully in accordance with all the observations of past and of present times, consider what it has cost of intellectual study. Copernicus, Galileo, Kepler, Euler, Lagrange, Laplace, all the great names which have exalted the character of man, by carrying out trains of reasoning unparalleled in every other science; these, and a host of others, each of whom might have been the Newton of another field, have all labored to work out, the consequences which resulted from that single law which *he* discovered. All that the human mind has produced—the brightest in genius, the most persevering in application, has been lavished on the details of the law of gravity.

Charles Babbage, English Inventor/Mathematician (1792–1871)

31.8 It [space travel] will free man from his remaining chains, the chains of gravity which still tie him to this planet. It will open to him the gates of heaven.

Wernher von Braun, German/American Rocket Engineer (1912–1977)

31.9 As soon as matter took over, the force of Newtonian gravity, which represents one of the most important characteristics of "ponderable" matter, came into play.

Werner Karl Heisenberg, German Physicist (1948)

31.10 You give a little push and bound through the air and come down and push off again. It's a very pleasant feeling to go loping across the surface. [Describing weightlessness during a moon walk.]

Commander Edgar D. Mitchell, Jr., American Astronaut (1971)

31.11 The force of gravity—though it is the first force with which we are acquainted, and though it is always with us, and though it is the one with a strength we most thoroughly appreciate—*is by far the weakest known force in nature*. It is first and rearmost.

Isaac Asimov, American Biochemist/Author (1976)

32. Hypothesis

A hypothesis may be simply defined as a guess. A scientific hypothesis is an intelligent guess.

ISAAC ASIMOV

32.1 I do not profess to be able thus to account for all the [planetary] motions at the same time; but I shall show that each by itself is well explained by its proper hypothesis.

Ptolemy, Egyptian Mathematician/Astronomer (circa 100 A.D.)

32.2 Therefore let us permit these new hypotheses to make a public appearance among old ones which are themselves no more probable, especially since they are wonderful and easy and bring with them a vast storehouse of learned observations.

Nicholas Copernicus, Polish Astronomer (1473–1543)

32.3 Hypotheses non fingo.
[I make no hypotheses.]

Isaac Newton, English Physicist/Mathematician (1642–1727)

32.4 Whatever things are not derived from objects themselves, whether by the external senses or by the sensation of internal thoughts, are to be taken as hypotheses. . . . Those things which follow from the phenomena neither by demonstration nor by the argument of induction, I hold as hypotheses.

Isaac Newton, English Physicist/Mathematician (1642–1727)

32.5 Human science is uncertain guess.

Matthew Prior, English Poet/Diplomat (1664–1721)

32.6 I have the result but I do not yet know how to get it.
Karl Friedrich Gauss, German Mathematician (1777–1855)

32.7 When a hypothesis has come to be born in the mind, or gained footing there, it leads a life so far comparable with the life of an organism, as that it assimilates matter from the outside world only when it is like in kind with it and beneficial; and when on the other hand, such matter is not like in kind but hurtful, the hypothesis, equally with the organism, throws it off, or, if forced to take it, gets rid of it again entirely.
Arthur Schopenhauer, German Philosopher (1788–1860)

32.8 I have steadily endeavored to keep my mind free so as to give up any hypothesis, however much beloved (and I cannot resist forming one on every subject) as soon as the facts are shown to be opposed to it.
Charles Robert Darwin, English Naturalist/Evolutionist (1809–1882)

32.9 An idea, like a ghost (according to the common notion of a ghost), must be spoken to a little before it will explain itself.
Charles Dickens, English Novelist (1812–1870)

32.10 We must never be too absorbed by the thought we are pursuing.
Claude Bernard, French Physiologist (1813–1878)

32.11 If an idea presents itself to us, we must not reject it simply because it does not agree with the logical deductions of a reigning theory.
Claude Bernard, French Physiologist (1813–1878)

32.12 The great tragedy of Science—the slaying of a beautiful hypothesis by an ugly fact.
Thomas Henry Huxley, English Biologist/Evolutionist (1825–1895)

32.13 A science cannot be played with. If an hypothesis is advanced that obviously brings into direct sequence of cause and effect all the phenomena of human history, we must accept it, and if we accept it, we must teach it.
Henry (Brooks) Adams, American Author/Historian (1838–1918)

32.14 System, moreover, dictates the scientists' hypotheses themselves: those are most welcome which are seen to conduce most to simplicity in the overall theory. Predictions, once they have been deduced from hypotheses, are subject to the discipline of evidence in turn; but the hypotheses have, at the time of hypothesis, only the considerations of systematic simplicity to recommend them.
Willard van Orman Quine, American Philosopher (1908–)

32.15 In science the primary duty of ideas is to be useful and interesting even more than to be "true."

Wilfred Trotter, English Philosopher/Scientist (1941)

32.16 The number of rational hypotheses that can explain any given phenomenon is infinite.

Robert Pirsig, American Author (1974)

33.
Invention

The greatest inventors are unknown to us.

Someone invented the wheel—but who?

ISAAC ASIMOV

33.1 Deus ex machina.
[The God from the machine.]
Lucian, Greek Rhetorician (120–180)

33.2 I, your servant, as I have pondered on the problem of the mill, have found, with the Lord's help, that by a device of art, one could construct a mill which works without water and by the wind alone; this could be constructed with less difficulty than a mill in the sea. This mill is not only easier for the crew to handle, but it will also work wherever it stands.
Leonardo da Vinci, Italian Architect/Artist/Inventor (1452–1519)

33.3 The industry of artificers maketh some small improvement of things invented; and chance sometimes in experimenting maketh us to stumble upon somewhat which is new; but all the disputation of the learned never brought to light one effect of nature before unknown.
Francis Bacon, English Philosopher/Essayist/Statesman (1561–1626)

33.4 The art of invention grows young with the things invented.
Francis Bacon, English Philosopher/Essayist/Statesman (1561–1626)

33.5 Th'invention all admir'd, and each, how he
To be th'inventor miss'd; so easy it seem'd,
Once found, which yet unfound most would have thought
Impossible.
John Milton, English Poet (1608–1674)

33.6 The greatest inventions were produced in times of ignorance; as the use of the compass, gunpowder, and printing; and by the dullest nation, as the Germans.

Jonathan Swift, Irish Satirist/Clergyman (1667–1745)

33.7 Want is the mistress of invention.

Susannah Centilivre, English Dramatist/Actress (1667–1723)

33.8 Invention is nothing more than a fine deviation from, or enlargement on a fine model. . . . Imitation, if noble and general, insures the best hope of originality.

Edward Bulwer-Lytton, English Politician/Author (1803–1873)

33.9 Only an inventor knows how to borrow, and every man is or should be an inventor.

Ralph Waldo Emerson, American Essayist/Philosopher/Poet (1803–1882)

33.10 It is frivolous to fix pedantically the date of particular inventions. They have all been invented over and over fifty times. Man is the arch machine, of which all these shifts drawn from himself are toy models. He helps himself on each emergency by copying or duplicating his own structure, just so far as the need is.

Ralph Waldo Emerson, American Essayist/Philosopher/Poet (1803–1882)

33.11 Invention breeds invention.

Ralph Waldo Emerson, American Essayist/Philosopher/Poet (1803–1882)

33.12 He that invents a machine augments the power of a man and the well-being of mankind.

Henry Ward Beecher, American Preacher/Author (1813–1887)

33.13 Galileo called doubt the father of invention; it is certainly the pioneer.

Christian Nestell Bovee, American Author (1820–1904)

33.14 In the present state of our knowledge it cannot be done; but I cannot say that it will always remain impossible, nor set the man down as mad who seeks to do it.

[To Mme. Daguerre regarding her concern that her husband was working in vain to invent the daguerreotype.]

Jean-Baptiste Dumas, French Chemist/Statesman (1827)

33.15 What hath God wrought?

[First telegraph message sent from the Supreme Court room in the Capitol to Baltimore.]

Samuel Morse, American Painter/Inventor (1832)

33.16 Name the greatest of all inventors: Accident.
Mark Twain (Samuel Langhorne Clemens), American Journalist/ Novelist (1835–1910)

33.17 A stone arrowhead is as convincing as a steam-engine.
Henry (Brooks) Adams, American Author/Historian (1838–1918)

33.18 Great discoveries and improvements invariably involve the cooperation of many minds. I may be given credit for having blazed the trail but when I look at the subsequent developments I feel the credit is due to others rather than to myself.
Alexander Graham Bell, Scottish/American Scientist/Inventor (1847–1922)

33.19 Mr. Watson, please come here. I want you.
[First words spoken over a telephone.]
Alexander Graham Bell, Scottish/American Scientist/Inventor (1876)

33.20 "Mary had a little lamb, whose fleece was white as snow!"
[First words from the phonograph.]
Thomas Alva Edison, American Inventor (1877)

33.21 A minor invention every ten days, and a big one every six months or so.
[His expectations for production at his new laboratory.]
Thomas Alva Edison, American Inventor (1847–1931)

33.22 Say I have lost all faith in patents, judges, and everything relating to patents.
Thomas Alva Edison, American Inventor (1847–1931)

33.23 Tell Selden to take his patent and go to hell with it.
[Referring to George Baldwin Selden's patent of an internal combustion engine modeled after George Brayton's. Ford fought Selden's monopoly in court for eight years—winning in 1911.]
Henry Ford, American Industrialist/Auto Maker (1863–1947)

33.24 I will build a motor car for the great multitude . . . so low in price that no man . . . will be unable to own one—and enjoy with his family the blessing of pleasure in God's great open spaces.
Henry Ford, American Industrialist/Auto Maker (1863–1947)

33.25 Civilization, a much abused word, stands for a high matter quite apart from telephones and electric lights.
Edith Hamilton, American Author (1867–1963)

33.26 The great object is to find the theory of the matter [of X-rays] before anyone else, for nearly every professor in Europe is now on the warpath.
Ernest Rutherford, British Physicist (1871–1937)

33.27 The utility of the typewriter is so great, its success so marked, its applications so numerous, that no prophetic vision is required to perceive that, ere long, it will become spread throughout the civilized world, like the clock and the sewing machine.

***Scientific American* (1886)**

33.28 An inventor is simply a fellow who doesn't take his education too seriously.

Charles F. Kettering, American Engineer/Inventor (1894–1971)

33.29 Besides electrical engineering theory of the transmission of messages, there is a larger field [cybernetics] which includes not only the study of language but the study of messages as a means of controlling machinery and society, the development of computing machines and other such automata, certain reflections upon psychology and the nervous system, and a tentative new theory of scientific method.

Norbert Wiener, American Inventor/Mathematician (1894–1964)

33.30 Once you ask the question, where is the Carbon-14, and where does it go, it's like one, two, three, you have [radiocarbon] dating.

Willard F. Libby, American Chemist (1908–1980)

33.31 What have you gentlemen done with my child [radio]? He was conceived as a potent instrumentality for culture, fine music, the uplifting of America's mass intelligence. You have debased this child, you have sent him out in the streets . . . to collect money from all and sundry, for hubba hubba and audio jitterbug. You have made of him a laughingstock to the intelligence, surely a stench in the nostrils of the gods of the ionosphere; you have cut time into tiny parcels called spots (more rightly "stains") wherewith the occasional fine program is periodically smeared with impudent insistence to buy or try. . . .

Some day the program director will attain the intelligent skill of the engineers who erected his towers and built the marvel he now so ineptly uses.

Lee De Forest, American Electrical Engineer/Inventor (1930)

33.32 Holography has seemed for years to epitomize an invention looking for a necessity.

***Scientific American* (1986)**

34. Laboratory

A neat and orderly laboratory is unlikely. It is, after all, so much a place of false starts and multiple attempts.

ISAAC ASIMOV

34.1 The bottom of the sea is the great laboratory.
John Playfair, Scottish Mathematician/Geologist (1748–1819)

34.2 [Laboratory:] a location for carrying out chemical, pharmaceutical, and biological experiments.
Maximilien-Paul-Emile Littré, French Lexicographer (1801–1881)

34.3 Put off your imagination, as you put off your overcoat, when you enter the laboratory. But put it on again, as you put on your overcoat, when you leave.
Claude Bernard, French Physiologist (1813–1878)

34.4 Without laboratories men of science are soldiers without arms.
Louis Pasteur, French Chemist/Microbiologist (1822–1895)

34.5 Oh, the serene peace of the laboratories and the libraries.
Louis Pasteur, French Chemist/Microbiologist (1822–1895)

34.6 When Raulin, Pasteur's chief assistant, took up his duties, the laboratory beneath the cellar had been given back to the rats, its legitimate proprietors. A very small building on the Rue d'Ulm, built as a counterpart to the caretaker's lodge, had just been put at the disposal of Pasteur. It had been well-nigh impossible to construct an incubator, which was indispensable for studying fermentations. Pasteur had built one in the well of the stairs, but he could only go in on his knees. Nevertheless, I have seen him spending many long hours there, for it is in this tiny incubator that all the studies on spontaneous generation were made and where the thousands of vessels employed for his celebrated experiments were daily examined. The movement which revolutionized the science of physical man, in all its aspects, started in this tiny hovel which we would hesitate to use as a rabbit hutch.

Emile Duclaux, French Biochemist (1840–1904)

34.7 You know, I *am* sorry for the poor fellows that haven't got labs to work in.

Ernest Rutherford, British Physicist (1871–1937)

34.8 Edison's greatest invention was that of the industrial research laboratory.

Norbert Wiener, American Inventor/Mathematician (1894–1964)

34.9 Most of my professional life these days is verbal—I no longer have a laboratory.

B.F. Skinner, American Psychologist (1904–)

34.10 Is it not striking that it is the only place where people work which derives its name from the Latin word *laborare—to work?*

Pierre Lecomte du Nouy, French Essayist/Historian/Scientist (1966)

34.11 The laboratory . . . I cannot write this word without emotion. For me, as for all my colleagues, it evokes so many memories, so many events, emotions, hopes. The laboratory is not only the setting of our material life, the place where a great part of our existence has been spent, but is above all, the nucleus of our intellectual life and even, at times, of our sentimental life.

Pierre Lecomte du Nouy, French Essayist/Historian/Scientist (1966)

35.
Light

If we were blind for one day each year,

how we would enjoy the other three hundred

and sixty-four.

ISAAC ASIMOV

35.1 And the light shineth in the darkness; and the darkness comprehendeth it not.

The Bible (circa 325 A.D.)

35.2 So when light generates itself in one direction drawing matter with it, it produces local motion; and when the light within matter is sent out and what is outside is sent in, it produces qualitative change. From this it is clear that corporeal motion is a multiplicative power of light, and this is a corporeal and natural appetite.

Robert Grosseteste, English Bishop/Educator (1168–1253)

35.3 Ethereal, first of things, quintessence, pure.

John Milton, English Poet (1608–1674)

35.4 One may conceive light to spread successively, by spherical waves.

Christian Huygens, Dutch Astonomer/Physicist (1629–1695)

35.5 The first Man I saw was of meagre Aspect, with sooty Hands and Face, his Hair and Beard long, ragged, and singed in several Places. His Clothes, Shirt, and Skin were all of the same Color. He had been Eight Years upon a Project for extracting Sun-Beams out of Cucumbers, which were to be put into Vials hermetically sealed, and let out to warm the Air in raw inclement

Summers. He told me, he did not doubt in Eight Years more, that he should be able to supply the Governor's Gardens with Sunshine at a reasonable Rate.

Jonathan Swift, Irish Satirist/Clergyman (1667–1745)

35.6 Is not light grander than fire? It is the same element in the state of purity.

Thomas Carlyle, Scottish Historian/Philosopher (1795–1881)

35.7 I am here to support the assertion that light of every kind is itself an electrical phenomenon—the light of the sun, the light of a candle, the light of a glowworm.

Heinrich Rudolf Hertz, German Physicist (1857–1894)

35.8 Light is always propagated in empty space with a definite velocity, "c," which is independent of the state of motion of the emitting body.

Albert Einstein, German/American Physicist (1879–1955)

35.9 That he [Einstein] may sometimes have missed the target . . . as, for example, in his hypothesis of light quanta, cannot really be held against him. [From his letter proposing the young Einstein as a member of the Royal Prussian Academy of Science.]

Max Planck, German Physicist (1858–1947)

35.10 There was a young lady named Bright,
Who traveled much faster than light.
She started one day
In the relative way,
And returned on the previous night.

Anonymous (circa 1950)

36.

Linguistics

There is something particularly human

about using tools; the first and most

important tool being language.

ISAAC ASIMOV

36.1 So out of the ground the Lord God formed every beast of the field and every bird of the air, and brought them to the man to see what he would call them; and whatever the man called every living creature, that was its name. The man gave names to all cattle, and to the birds of the air, and to every beast of the field. . . .

***The Bible* (circa 725 B.C.)**

36.2 Now the whole earth had one language and few words. . . . Then they said, "Come, let us build ourselves a city, and a tower with its top in the heavens, and let us make a name for ourselves, lest we be scattered abroad upon the face of the whole earth." And the Lord came down to see the city and the tower, which the sons of men had built. And the Lord said, "Behold, they are one people, and they have all one language; and this is only the beginning of what they will do; and nothing that they propose to do will now be impossible for them. Come, let us go down, and there confuse their language, that they may not understand one another's speech." So the Lord scattered them abroad from there over the face of all the earth, and they left off building the city. Therefore its name was called Babel, because there the Lord confused the language of all the earth. . . .

***The Bible* (circa 725 B.C.)**

36.3 Let thy speech be short, comprehending much in few words.

***The Bible* (circa 725 B.C.)**

36.4 Speech is the representation of the mind, and writing is the representation of speech.
Aristotle, Greek Philosopher (384 B.C.–322 B.C.)

36.5 Usage, in which lies the decision, the law, and the norm of speech.
Horace (Quintus Horatius Flaccus), Roman Poet (65 B.C.–8 B.C.)

36.6 Because words pass away as soon as they strike upon the air, and last no longer than their sound, men have by means of letters formed signs of words. Thus the sounds of the voice are made visible to the eye, not of course as sounds, but by means of certain signs.
Saint Augustine of Hippo, North African Theologian (354–430)

36.7 If man were by nature a solitary animal, the passions of the soul by which he was conformed to things so as to have knowledge of them would be sufficient for him; but since he is by nature a political and social animal it was necessary that his conceptions be made known to others. This he does through vocal sound. Therefore there had to be significant vocal sounds in order that men might live together. Whence those who speak different languages find it difficult to live together in social unity.
Saint Thomas Aquinas, Italian Theologian/Scholar (1225–1274)

36.8 Man alone amongst the animals speaks and has gestures and expression which we call rational, because he alone has reason in him. And if anyone should say in contradiction that certain birds talk, as seems to be the case with some, especially the magpie and the parrot, and that certain beasts have expression or gestures, as the ape and some others seem to have, I answer that it is not true that they speak, nor that they have gestures, because they have no reason, from which these things need proceed; nor do they purpose to signify anything by them, but they merely reproduce what they see and hear.
Dante Alighieri, Italian Poet (1265–1321)

36.9 Ye knowe eek, that in forme of speche is chaunge
With-inne a thousand yeer, and wordes tho
That hadden prys, now wonder nyce and straunge
Us thinketh hem; and yet they spake hem so,
And spedde as wel in love as men now do.
Geoffrey Chaucer, English Poet (1340?–1400)

36.10 After a speech is fully fashioned to the common understanding, and accepted by consent of a whole country and nation, it is called a language.
George Puttenham, English Author/Critic (1530–1590)

36.11 This is your devoted friend, sir, the manifold linguist.
William Shakespeare, English Dramatist/Poet (1564–1616)

36.12 Cassius: Did Cicero say anything?
Casca: Ay, he spoke Greek.

Cassius: To what effect?
Casca: Nay, an' I tell you that I'll ne'er look you i' the face again; but those that understood him smiled at one another and shook their heads;
but, for mine own part, it was Greek to me.
William Shakespeare, English Dramatist/Poet (1564–1616)

36.13 Custom is the most certain mistress of language, as the public stamp makes the current money.
Ben Jonson, English Dramatist/Poet (1572–1637)

36.14 God loveth adverbs.
Bishop Joseph Hall, English Clergyman (1574–1656)

36.15 Syllables govern the world.
John Selden, English Jurist (1584–1654)

36.16 I trade both with the living and the dead for the enrichment of our native language.
John Dryden, English Poet/Critic/Dramatist (1631–1700)

36.17 God having designed man for a sociable creature, furnished him with language, which was to be the great instrument and cementer of society.
John Locke, English Philosopher (1632–1704)

36.18 The wisdom of nations lies in their proverbs, which are brief and pithy.
William Penn, American Politician (1644–1718)

36.19 Every living language, like the perspiring bodies of living creatures, is in perpetual motion and alteration; some words go off, and become obsolete; others are taken in, and by degrees grow into common use; or the same word is inverted to a new sense and notion, which in tract of time makes as observable a change in the air and features of a language as age makes in the lines and mien of a face.
Richard Bentley, English Clergyman/Author/Critic (1662–1742)

36.20 Our sons their fathers' failing language see.
Alexander Pope, English Poet/Satirist (1688–1744)

36.21 The less men think, the more they talk.
Baron de la Brède et de Montesquieu, French Philosopher/Author (1689–1755)

36.22 Write with the learned, pronounce with the vulgar.
Benjamin Franklin, American Inventor/Statesman (1706–1790)

36.23 Language is the only instrument of science, and words are but the signs of ideas.
Samuel Johnson, English Lexicographer/Poet/Critic (1709–1784)

36.24 Languages are the pedigrees of nations.
Samuel Johnson, English Lexicographer/Poet/Critic (1709–1784)

36.25 Academies have been instituted to guard the avenues of their languages, to retain fugitives, and repulse intruders; but their vigilance and activity have hitherto been vain; sounds are too volatile and subtile [sic] for legal restraints; to enchain syllables, and to lash the wind, are equally the undertakings of pride, unwilling to measure its desires by its strength.
Samuel Johnson, English Lexicographer/Poet/Critic (1709–1784)

36.26 Accent is the soul of a language; it gives feeling and truth to it.
Jean-Jacques Rousseau, Swiss/French Philosopher/Author (1712–1778)

36.27 Words learned by rote a parrot may rehearse,
But talking is not always to converse;
Not more distinct from harmony divine,
The constant creaking of a country sign.
William Cowper, English Poet (1731–1800)

36.28 He who is ignorant of foreign languages knows not his own.
Johann Wolfgang von Goethe, German Poet/Dramatist/Novelist (1749–1832)

36.29 Egad, I think the interpreter is the hardest to be understood of the two!
Richard Brinsley Sheridan, Irish Dramatist (1751–1816)

36.30 I was in a Printing-house in Hell, and saw the method in which knowledge is transmitted from generation to generation.
William Blake, English Poet/Painter/Engraver (1757–1827)

36.31 Language, as well as the faculty of speech, was the immediate gift of God.
[From the preface to his dictionary.]
Noah Webster, American Author/Lexicographer/Philologist (1758–1843)

36.32 Language is the expression of ideas, and if the people of one country cannot preserve an identity of ideas they cannot retain an identity of language.
Noah Webster, American Author/Lexicographer/Philologist (1758–1843)

36.33 Babylon,
Learned and wise, hath perished utterly,
Nor leaves her speech one word to aid the sigh
That would lament her.
William Wordsworth, English Poet (1770–1850)

36.34 Silence is one great art of conversation.
William Hazlitt, English Critic/Essayist (1778–1830)

36.35 Examine Language; what, if you except some few primitive elements (of natural sound), what is it all but Metaphors, recognized as such, or no longer recognized?

Thomas Carlyle, Scottish Historian/Philosopher (1795–1881)

36.36 The Romans would never have found time to conquer the world if they had been obliged first to learn Latin.

Heinrich Heine, German/Jewish Poet/Author (1797–1856)

36.37 Debate is masculine; conversation is feminine.

Amos Bronson Alcott, American Educator/Philosopher (1799–1888)

36.38 On the greatest and most useful of all human inventions, the invention of alphabetical writing, Plato did not look with much complacency. He seems to have thought that the use of letters had operated on the human mind as the use of the go-cart in learning to walk, or of corks in learning to swim, is said to operate on the human body. It was a support which, in his opinion, soon became indispensable to those who used it, which made vigorous exertion first unnecessary, and then impossible. The powers of the intellect would, he conceived, have been more fully developed without this delusive aid. Men would have been compelled to exercise the understanding and the memory, and, by deep and assiduous mediation, to make truth thoroughly their own. Now, on the contrary, much knowledge is traced on paper, but little is engraved on the soul.

Thomas Babington Macaulay, English Historian/Author/Statesman (1800–1859)

36.39 Language is the archives of history. . . . Language is fossil poetry.

Ralph Waldo Emerson, American Essayist/Philosopher/Poet (1803–1882)

36.40 Language is the city to the building of which every human being brought a stone.

Ralph Waldo Emerson, American Essayist/Philosopher/Poet (1803–1882)

36.41 The language of the street is always strong. What can describe the folly and emptiness of scolding like the word jawing?

Ralph Waldo Emerson, American Essayist/Philosopher/Poet (1803–1882)

36.42 Language is the amber in which a thousand precious and subtle thoughts have been safely imbedded and preserved.

Richard Chenevix Trench, English Poet/Philologist (1807–1886)

36.43 The knowledge of the ancient languages is mainly a luxury.

John Bright, English Statesman (1811–1889)

36.44 I believe that our Great Maker is preparing the world, in His own good time, to become one nation, speaking one language.

Ulysses S. Grant, American President (1822–1885)

36.45 No man fully capable of his own language ever masters another.

George Bernard Shaw, Irish Dramatist/Critic (1856–1950)

36.46 Words and magic were in the beginning one and the same thing, and even today words contain much of their magical power.

Sigmund Freud, Austrian Psychiatrist/Psychoanalyst (1856–1939)

36.47 Broadly speaking, the short words are the best, and the old words are the best of all.

Sir Winston Churchill, English Statesman/Author (1874–1965)

36.48 Sometimes I doubt whether there is divine justice; all parts of the human body get tired eventually—except the tongue. And I feel this is unjust.

Konrad Adenauer, German Statesman (1876–1967)

36.49 We shall never understand each other until we reduce the language to seven words.

Khalil Gibran, Lebanese Mystic/Poet (1883–1931)

36.50 We dissect nature along lines laid down by our native languages. The categories and types that we isolate from the world of phenomena we do not find there because they stare every observer in the face; on the contrary, the world is presented in a kaleidoscopic flux of impressions which has to be organized by our minds—and this means largely by the linguistic systems in our minds.

Benjamin Whorf, American Linguist (1897–1941)

36.51 The notion of reality independent of language is carried over by the scientist from his earliest impressions, but the facile reification of linguistic features is avoided or minimized.

Willard van Orman Quine, American Philosopher (1908–)

36.52 I think that in order to achieve progress in the study of language and human cognitive faculties in general it is necessary first to establish "psychic distance" from the "mental facts" to which Kohler referred, and then to explore the possibilities for developing explanatory theories. . . . We must recognize that even the most familiar phenomena require explanation and that we have no privileged access to the underlying mechanisms, no more so than in physiology or physics.

Noam Chomsky, American Linguist/Philosopher (1928–)

36.53 It is quite natural to expect that a concern for language will remain central to the study of human nature, as it has been in the past. Anyone concerned with the study of human nature and human capacities must somehow come to grips with the fact that all normal humans acquire language. . . .

Noam Chomsky, American Linguist/Philosopher (1928–)

36.54 Language acquires life and historically evolves precisely here, in concrete verbal communication, and not in the abstract linguistic system of language forms, nor in the individual psyches of speakers.

M.M. Bakhtin and V.N. Yoloshinov, Russian Linguists (1929)

36.55 The terms "Aryan" and "Semitic" have no racial significance whatsoever. They simply denote linguistic families.

American Anthropological Association, Resolution (1938)

36.56 The average working-class Philadelphian doesn't see anything wrong with Philadelphia speech, but in New York City the average working-class person feels there's something wrong with the way he talks.

William Labov, American Linguist/Educator (1971)

36.57 At Marine World, we're working with computers to develop a human-dolphin code, analogous to the Morse code used in telegraphy. The project is called JANUS—for Joint Analog Numerical Understanding System. Like the Roman god Janus, it has two "faces"—a dolphin side and a human side.

John Lilly, American Neurophysiologist/Spiritualist (1984)

36.58 We do not communicate by transmitting signs or signals, but by creating and manipulating social situations. *Communication is the creation of community.*

Jay Lemke, American Anthropologist (1986)

37. Logic

When you say, "The burned child dreads the fire," you mean that he is already a master of induction.

ISAAC ASIMOV

37.1 The only hope of science is genuine induction.
Francis Bacon, English Philosopher/Essayist/Statesman (1561–1626)

37.2 Men are rather beholden . . . generally to chance, or anything else, than to logic, for the invention of arts and sciences.
Francis Bacon, English Philosopher/Essayist/Statesman (1561–1626)

37.3 Deduction, which takes us from the general proposition to facts again—teaches us, if I may so say, to anticipate from the ticket what is inside the bundle.
Thomas Henry Huxley, English Biologist/Evolutionist (1825–1895)

37.4 Discovery should come as an adventure rather than as the result of a logical process of thought. Sharp, prolonged thinking is necessary that we may keep on the chosen road, but it does not necessarily lead to discovery.
Theobald Smith, American Pathologist (1859–1934)

37.5 The distinctive Western character begins with the Greeks, who invented the habit of deductive reasoning and the science of geometry.
Bertrand Russell, English Philosopher/Mathematician (1872–1970)

37.6 There is no inductive method which could lead to the fundamental concepts of physics. Failure to understand this fact constituted the basic philosophical error of so many investigators of the nineteenth century.
Albert Einstein, German/American Physicist (1879–1955)

37.7 Induction is the process of generalizing from our known and limited experience, and framing wider rules for the future than we have been able to test fully. At its simplest, then, an induction is a habit or an adaptation—the habit of expecting tomorrow's weather to be like today's, the adaptation to the unwritten conventions of community life.

Jacob Bronowski, Polish/British Mathematician/Science Writer (1908–1974)

37.8 It is time, therefore, to abandon the superstition that natural science cannot be regarded as logically respectable until philosophers have solved the problem of induction. The problem of induction is, roughly speaking, the problem of finding a way to prove that certain empirical generalizations which are derived from past experience will hold good also in the future.

A.J. Ayer, British Philosopher (1910–)

37.9 Among the obstacles to scientific progress a high place must certainly be assigned to the analysis of scientific procedure, which logic has provided. . . .

F.C.S. Schiller, German Philosopher (1917–)

38.
Marine Biology

Life originated in the sea, and about

eighty percent of it is still there.

ISAAC ASIMOV

38.1 So God created the great sea monsters and every living creature that moves, with which the waters swarm, according to their kinds.
The Bible (circa 725 B.C.)

38.2 The monsters of the sea moved continually hither and thither, and the wild beasts swim among the sluggish and slowly creeping ships.
Himlico of Carthage, Phoenician Navigator (circa 500 B.C.)

38.3 Third Fisherman: Master, I marvel how the fishes live in the sea.
First Fisherman: Why, as men do a-land; the great ones eat up the little ones: I can compare our rich misers to nothing so fitly as to a whale; a' plays and tumbles, driving the poor fry before him, and at last devours them all at a mouthful: such whales have I heard on o' the land, who never leave gaping till they've swallowed the whole parish, church, steeple, bells, and all.
William Shakespeare, English Dramatist/Poet (1564–1616)

38.4 The pleasant'st angling is to see the fish
Cut with her golden oars the silver stream,
And greedily devour the treacherous bait.
William Shakespeare, English Dramatist/Poet (1564–1616)

38.5 Two canons have just arrived here from Marseilles in order to recount to the Pope that in the Sea of Pre there is to be observed such a conglomeration of dolphins that not only do they interfere with the fishing, a most valuable source of revenue, but also with ships sailing on the sea. A Papal Brief has been bestowed . . . in which it is ordered that the matter should be taken in hard by means of prayers in the churches, processions, and much fasting. Moreover, the Pope anathematizes this vermin, so that, with God's help, it may perish.

Anonymous, Business Agent for the Fugger Banking Family (1589)

38.6 And an ingenious Spaniard says, that rivers and the inhabitants of the watery element were made for wise men to contemplate, and fools to pass by without consideration.

Izaak Walton, English Writer (1593–1683)

38.7 The King was then with no common satisfaction, expressing his desire in no particular to have the stellar fish engraven and printed; We wish very much, that you could procure a particular description of the said fish viz: whether it be common there; what is observable in it when alive; what color it hath then; what kind of motion in water; what use it maketh of all that curious workmanship which nature hath adorned it with?
[Letter to John Winthrop, Jr., March 26, 1670, about specimens sent by Winthrop to the Society.]

Henry Oldenburg, Secretary, Royal Society of London (1615–1677)

38.8 Amphibious Animals link the Terrestrial and Aquatique together; Seals live at Land and at Sea, and Porpoises have the warm Blood and Entrails of a Hog, not to mention what is confidently reported of Mermaids or Sea-men.

Joseph Addison, English Essayist/Poet/Politician (1672–1719)

38.9 Our plenteous streams a various race supply,
The bright-eye perch with fins of Tyrian dye,
The silver eel, in shining volumes roll'd,
The yellow carp, in scales bedropp'd with gold,
Swift trouts, diversified with crimson stains,
And pikes, the tyrants of the tyrants of the wat'ry plains.

Alexander Pope, English Poet/Satirist (1688–1744)

38.10 A whale!
Down it goes, and more and more
up goes its tail!

Taniguchi Buson, Japanese Poet/Painter (1715–1783)

38.11 The very deep did rot: O Christ!
That ever this should be!
Yea, slimy things did crawl with legs
Upon the slimy sea.

Samuel Taylor Coleridge, English Poet/Critic (1772–1834)

38.12 The whale that wanders round the Pole
Is not a table fish.
You cannot bake or boil him whole
Nor serve him in a dish.

Hilaire Belloc, English Author (1870–1953)

38.13 The greatest wealth of the sea is not to be found in its mineral resources, but in its natural productivity, in plants as well as animals.

Claude E. Zobell, American Marine Biologist/Bacteriologist (1944)

38.14 No philosopher's or poet's fancy, no myth of a primitive people has ever exaggerated the importance, the usefulness, and above all the marvelous beneficence of the ocean for the community of living things.

L.J. Henderson, American Ecologist (1960)

38.15 The frillshark has many anatomical features similar to those of the ancient sharks that lived 25 to 30 million years ago. It has too many gills and too few dorsal fins for a modern shark, and its teeth, like those of fossil sharks, are three-pronged and briarlike. Some ichthyologists regard it as a relic derived from very ancient shark ancestors that have died out in the upper waters but, through this single species, are still carrying on their struggle for earthly survival, in the quiet of the deep sea.

Rachel Carson, American Marine Biologist/Author (1961)

38.16 It is only at the beginning of the age of the dinosaurs that the deep sea, hitherto bare of organisms, was finally invaded by life.

Isaac Asimov, American Biochemist/Author (1965)

38.17 Few organisms have been as well known to the layman but as poorly known to science as the chambered nautilus.

W. Bruce Saunders, American Paleontologist (1986)

38.18 They want to punish us because of the turtles. But here we don't care about the turtles. Since man appeared on the earth he has destroyed whatever got in his way.

Andreas Logothetis, Greek Mayor (1986)

38.19 Ironically, for an organism so persistent in evolutionary time, a horseshoe crab cannot right itself if flipped by the surf. This happens to many crabs in each tide cycle and only a few of these will be righted by wave action in a later tide. Perhaps to compensate for this curious inability, a horseshoe crab can withstand many days of exposure as long as its book lungs remain damp.

J.P. Myers, American Zoologist/Ornithologist (1986)

38.20 The whales like to play with stones and seaweed. A male beluga dives and emerges with a long frond of seaweed swirling about its head. Other whales instantly rush at him, and there is a great deal of bumping and squeaking as they tear his trophy into tatters. A whale may swim around for hours with a plate-sized stone held firmly in its mouth. And often a whale will surface with a stone balanced skillfully but precariously atop its head. . . . The moment other whales spot the stone carrier, they bump and jostle him until the stone falls off.

Fred Bruemmer, American Naturalist/Photographer (1986)

39.
Matter

Matter is concentrated energy. When a small bit of it is converted to other forms of energy, the result is an H-bomb.

ISAAC ASIMOV

39.1 Our view is that there is a matter of the perceptible bodies, but that this is not separable but is always together with a contrariety, from which the so-called "elements" come to be.

Aristotle, Greek Philosopher (384 B.C.– 322 B.C.)

39.2 It is not possible for form to do without matter because it is not separable, nor can matter itself be purged of form.

Robert Grosseteste, English Bishop/Educator (1168–1253)

39.3 We should note that prime matter, and even form, are neither generated nor corrupted, inasmuch as every generation is from something to something. That from which generation arises is matter; that to which it proceeds is form. If, therefore, matter and form were generated, there would have to be a matter of matter and a form of form *ad infinitum*. Hence, properly speaking, only composites are generated.

Saint Thomas Aquinas, Italian Theologian/Scholar (1225–1274)

39.4 But if any skillful minister of nature shall apply force to matter, and by design torture and vex it, in order to [effect] its annihilation, it, on the contrary being brought under this necessity, changes and transforms itself into a strange variety of shapes and appearances; for nothing but the power of the Creator can annihilate, or truly destroy it.

Francis Bacon, English Philosopher/Essayist/Statesman (1561–1626)

39.5 The greatest inventions are those inquiries which tend to increase the power of man over matter.

Benjamin Franklin, American Inventor/Statesman (1706–1790)

39.6 For in the concept of matter I do not think [of] its permanence, but only its presence in the space which it occupies.

Immanuel Kant, German Philosopher (1724–1804)

39.7 There are three distinctions in the kinds of bodies, or three states, which have more especially claimed the attention of philosophical chemists; namely, those which are marked by the terms *elastic fluids, liquids,* and *solids.*

John Dalton, English Physicist/Chemist (1766–1844)

39.8 The first step toward numerical reckoning of the properties of matter is the discovery of a continuously varying action of some kind, and the means of measuring it in terms of some arbitrary unit or scale division.

William Thomson, Lord Kelvin, British Physicist (1824–1907)

39.9 We must regard it rather as an accident that the Earth (and presumably the whole solar system) contains a preponderance of negative electrons and positive protons. It is quite possible that for some of the stars it is the other way about.

Paul A.M. Dirac, British Physicist (1902–)

39.10 Naturally, some intriguing thoughts arise from the discovery that the three chief particles making up matter—the proton, the neutron, and the electron—all have antiparticles. Were particles and antiparticles created in equal numbers at the beginning of the universe? If so, does the universe contain worlds, remote from ours, which are made up of antiparticles?

Isaac Asimov, American Biochemist/Author (1965)

39.11 If there were not something of mind in matter, how could matter change the mind?

Albert Hofmann, Swiss Chemist (1984)

39.12 There is already overwhelming evidence that the visible matter within galaxies may account for less than 10 percent of the galaxies' actual mass: the rest, not yet directly detectable by observers on the earth, is probably distributed within and around each galaxy.

Lawrence M. Krauss, American Physicist/Astronomer (1986)

39.13 The physical structure of matter as we experience it does not permit the existence of fields like those we have found.
[Discussing a magnetic field of 700 million gauss found in the white dwarf PG 1031+234.]

Gary D. Schmidt, American Astronomer (1986)

39.14 The real question in everybody's mind is why the universe seems to be made of matter when, on a cosmic scale, antimatter is just as easy to make. It's one of the outstanding big mysteries.

Robert L. Forward, American Engineer (1987)

39.15 Even one antihelium particle would be clear proof that there are large aggregations of antimatter out there. That gets very exciting.

Martin H. Israel, American Astrophysicist (1987)

39.16 In theory, whole islands of antimatter could be floating in the universe, cut off from matter by the empty void of space. If a large chunk of antimatter fell to Earth, the planet would be vaporized in a blinding flash of energy.

William J. Broad, American Journalist (1987)

40.
Medicine

Such is the respect for physicians that most people are astonished when one of them falls sick—and yet they do.

ISAAC ASIMOV

40.1 If thou examinest a man having a break in the column of his nose, his nose being disfigured, and a [depression] being in it, while the swelling that is on it protrudes, [and] he had discharged blood from both his nostrils, thou shouldst say concerning him: "One having a break in the column of his nose. An ailment which I will treat."

Thou shouldst cleanse [it] for him [with] two plugs of linen. Thou shouldst place two [other] plugs of linen saturated with grease in the inside of his two nostrils. Thou shouldst put [him] at his mooring stakes until the swelling is drawn out. Thou shouldst apply for him stiff rolls of linen by which his nose is held fast. Thou shouldst treat him afterward [with] lint, every day until he recovers.

[From an ancient Egyptian document called "The Edwin Smith Surgical Papyrus"—the earliest known document in the history of science.]

Anonymous (circa 1700 B.C.)

40.2 A physician ought to have his shop provided with plenty of all necessary things, as lint, rollers, splinters: let there be likewise in readiness at all times another small cabinet of such things as may serve for occasions of going far from home; let him have also all sorts of plasters, potions, and purging medicines, so contrived that they may keep some considerable time, and likewise such as may be had and used whilst they are fresh.

Hippocrates, Greek Physician (460 B.C.?–377 B.C.?)

40.3 I swear by Apollo Physician, by Asclepiades, by Health, by Panacea, and all the gods and goddesses, making them my witnesses, that according to my ability and judgement I will carry out this oath. . . .

To hold my instructor in this art equal to my own parents; to make him partner in my livelihood; when he is in need of money to share mine with him; to treat his family as my own brothers, and to teach them this art if they want to learn, without charge or indenture. . . .

I will give no lethal medicine to anyone if asked, nor suggest any such counsel. . . . I will use treatment to help the sick according to my skill and judgement but never with a view of injury or wrongdoing. . . .

Into whatever houses I enter I will proceed for the benefit of the sick, and will abstain from every voluntary act of mischief or corruption, further, from the seduction of females and males, of freemen and slaves. . . .

Whatever, in connection with my professional practice, or not in connection with it, I see or hear, in the life of men, which should not be spoken abroad, I will not divulge, believing that all such things should be kept secret. . . .

Hippocrates, Greek Physician (460 B.C.?–377 B.C.?)

40.4 To do nothing is sometimes a good remedy.

Hippocrates, Greek Physician (460 B.C.?–377 B.C.?)

40.5 He will manage the cure best who foresees what is to happen from the present condition of the patient.

Hippocrates, Greek Physician (460 B.C.?–377 B.C.?)

40.6 Some remedies are worse than the disease.

Publius Syrus, Roman Poet (50?–100)

40.7 While the Kings were quarreling with each other again and once more making preparations for civil war, dysentery spread throughout the whole of Gaul. Those who caught it had a high temperature, with vomiting and severe pains in the small of the back: their heads ached and so did their necks. The matter they vomited up was yellow or even green. Many people maintained that some secret poison must be the cause of this. The country-folk imagined that they had boils inside their bodies; and actually this is not as silly as it sounds, for as soon as cupping-glasses were applied to their shoulders or legs, great tumors formed, and when these burst and discharged their pus, they were cured. Many recovered their health by drinking herbs which are known to be antidotes to poisons.

Gregory of Tours, Frankish Ecclesiastic/Historian (538–594)

40.8 At the same time died Austrechild, King Guntram's Queen, and of the same disease. . . . As Herod had done before her, she is said to have made this last request to the King: "I should still have some hope of recovery if my death had not been made inevitable by the treatment prescribed for me by these wicked doctors. It is the medicines which they have given me which have robbed me of my life and forced me thus to lose the light of day. I beseech you, do not let me die unavenged. Give me your solemn word, I beg you, that you will cut their throats the moment that my eyes have closed in death."

Gregory of Tours, Frankish Ecclesiastic/Historian (538–594)

40.9 A citizen of Tours called Wistrimund, commonly known as Tatto, was suffering terribly from toothache. His whole jaw was swollen. He went off to find this saintly man. Aredius placed his hand on the bad place, whereupon the pain immediately stopped and never plagued the man again.

Gregory of Tours, Frankish Ecclesiastic/Historian (538–594)

40.10 In fact, almost everything in this isle [Ireland] confers immunity to poison, and I have seen that folk suffering from snake-bite have drunk water in which scrapings from the leaves of books from Ireland had been steeped, and that this remedy checked the spreading poison and reduced the swelling.

Bede, English Monk/Scholar (673?–735)

40.11 And therefore, sir, as you desire to live,
A day or two before your laxative,
Take just three worms, nor under nor above,
Because the gods unequal numbers love.
These digestives prepare you for your purge,
Of fumetery, centaury, and spurge;
And of ground-ivy add a leaf or two.
All which within our yard or garden grow.
Eat these, and be, my lord, of better cheer:
Your father's son was never born to fear.

Geoffrey Chaucer, English Poet (1340?–1400)

40.12 Strive to preserve your health; and in this you will the better succeed in proportion as you keep clear of the physicians, for their drugs are a kind of alchemy concerning which there are no fewer books than there are medicines.

Leonardo da Vinci, Italian Architect/Artist/Inventor (1452–1519)

40.13 Those generally run through the whole course of life assigned to them by Heaven who do not derange their bodily organs, but keep them in their proper state; or if they change, it is for the better, and not for the worse.

Niccolò Machiavelli, Italian Statesman/Political Philosopher (1469–1527)

40.14 The Indians of our Occidental Indies use the tobacco to take away the weariness and to take lightsomeness of their labor, for in their dances they be so much wearied, they remain so weary, that they can scarcely stir. And so that they may labor the next day and return to that foolish exercise, they take at the mouth and nose the smoke of the tobacco and they remain as dead people. And being so, they be eased in that sort that when they be awakened of their sleep, they remain without weariness and may return to their labor as much more, and so they do always when they have need of it. For with that sleep they receive their strength and be much the lustier.

Nicolas Monardes, Spanish Physician (1493–1588)

40.15 So, then, the Tincture of the Philosophers is a universal medicine, and consumes all diseases, by whatsoever name they are called, just like an invisible fire. The dose is very small, but its effect is most powerful. By means thereof I have cured the leprosy, venereal disease, dropsy, the falling sickness, colic, scab, and similar afflictions; also lupus, cancer, noli-metangere, fistulas, and the whole race of internal diseases, more surely than one could believe.

Paracelsus (Theophrastus Bombastius), Swiss/German Chemist/ Physician (1493–1541)

40.16 That physician will hardly be thought very careful of the health of others who neglects his own.

François Rabelais, French Monk/Physician/Satirist (1494?–1553?)

40.17 Men today who have had an irreproachable training in the art are seen to abstain from the use of the hand as from the plague, and for this very reason, lest they should be slandered by the masters of the profession as barbers. . . . For it is indeed above all things the wide prevalence of this hateful error that prevents us even in our age from taking up the healing art as a whole, makes us confine ourselves merely to the treatment of internal complaints, and, if I may utter the blunt truth once for all, causes us, to the great detriment of mankind, to study to be healers only in a very limited degree.

Andreas Vesalius, Belgian Physician/Anatomist (1514–1564)

40.18 I treated him—God cured him.

Ambroise Paré, French Surgeon (1517–1590)

40.19 A healthy body is a guest-chamber for the soul; a sick body is a prison.

Francis Bacon, English Philosopher/Essayist/Statesman (1561–1626)

40.20 If the just cure of a disease be full of peril, let the physician resort to palliation.

Francis Bacon, English Philosopher/Essayist/Statesman (1561–1626)

40.21 But that which was most sadd & lamentable was, that in 2. or 3. moneths time halfe of their company dyed, espetialy in Jan: & February, being the depth of winter, and wanting houses & other comforts; being infected with the scurvie & other diseases, which this long vioage & their inacomodate condition had brought upon them; so as ther dyed some times 2. or 3. of a day, in the foresaid time; that of 100. & odd persons, scarce 50. remained. And of these in the time of most distres, ther was but 6. or 7. sound persons, who, to their great comendations be it spoken, spared no pains, night nor day, but with abundance of toyle and hazard of their owne health, fetched them woode, made them fires, drest them meat, made thair beads, washed their lothsome cloaths, cloathed & uncloathed them; in a word, did all the homly and necessarie offices for them which dainty and quesie stomacks cannot endure to hear named. [Discussing the winter of 1620–1621.]

William Bradford, American Governor (1590–1657)

40.22 To usher in again old Janus, I send you a parcel of Indian perfume, which the Spaniard calls the holy herb, in regard of the various virtues it hath, but we call it tobacco. . . . If moderately and seasonably taken . . . 'tis good for many things; it helps digestion taken a while after meat, it makes one void rheum, breaks wind, and it keeps the body open. A leaf or two being steeped o'er night in a little white wine is a vomit that never fails in its operation. . . . The smoke of it is one of the wholesomest scents that is against all contagious airs, for it o'ermasters all other smells. . . . It is good to fortify and preserve the sight, the smoke being let in round about the balls of the eyes once a week, and frees them from all rheums, driving them back by way of repercussion. Being taken backward, it is excellent good against the colic, and taken into the stomach, 'twill heat and clense it.

James Howell, Welsh Author (1594–1666)

40.23 You are still sending to the apothecaries and still crying out to fetch Master Doctor to me; but our apothecary's shop is our garden and our doctor a good clove of garlic.

Anonymous (1608)

40.24 There are affairs and diseases which remedies aggravate at certain times; and great ability is shown by knowing when it is dangerous to make use of them.

François de la Rochefoucauld, French Writer/Moralist (1613–1680)

40.25 It is a wearisome disease to preserve health by too strict a regimen.

François de la Rochefoucauld, French Writer/Moralist (1613–1680)

40.26 March 24, 1672. I saw the surgeon cut off the leg of a wounded sailor, the stout and gallant man enduring it with incredible patience without being bound to his chair as usual on such painful occasions. I had hardly courage enough to be present. Not being cut off high enough, the gangrene prevailed, and the second operation cost the poor creature his life.

John Evelyn, English Diarist (1620–1706)

40.27 How simple the beginnings of this art were, may be observed by the story of Aesculapius going about the country with a dog and a she-goat always following, both which he used much in his cures; the first for licking all ulcerated wounds, and the goat's milk for diseases of the stomach and lungs.

William Temple, English Statesman/Essayist (1628–1699)

40.28 Among the plants of our soil and climate, those I esteem of greatest virtue and most friendly to health are sage, rue, saffron, alehoof, garlic, and elder.

William Temple, English Statesman/Essayist (1628–1699)

40.29 Our physicians have observed that, in process of time, some diseases have abated of their virulence, and have, in a manner, worn out their malignity, so as to be no longer mortal.

John Dryden, English Poet/Critic/Dramatist (1631–1700)

40.30 Were it my business to understand physic, would not the safe way be to consult nature herself in the history of diseases and their cures, than espouse the principles of the dogmatists, methodists, or chemists?

John Locke, English Philosopher (1632–1704)

40.31 What Doctor is there, who while he treats a disease unknown to him, might be at ease, until he had clearly perceived the nature of this disease and its hidden causes?

Hermann Boerhaave, Dutch Physician (1668–1738)

40.32 A wealthy doctor who can help a poor man, and will not without a fee, has less sense of humanity than a poor ruffian who kills a rich man to supply his necessities. It is something monstrous to consider a man of a liberal education tearing out the bowels of a poor family by taking for a visit what would keep them a week.

Joseph Addison, English Essayist/Poet/Politician (1672–1719)

40.33 I have patiently born with abundance of Clamour and Ralary, for beginning a new Practice here (for the Good of the Publick) which comes well Recommended, from Gentlemen of Figure & Learning, and which well agrees to Reason, when try'd & duly considered, viz. Artificially giving the Small Pocks, by Inoculation, to One of my Children, and Two of my Slaves, in order to prevent the hazard of Life. . . . and they never took one grain or drop of Medicine since, & are perfectly well.
[About the public uproar over smallpox inoculation in Boston.]

Zabdiel Boylston, American Physician (1679–1766)

40.34 The art of causing intemperance and health to exist in the same body is as chimerical as the philosopher's stone, judicial astrology, and the theology of the magi.

François-Marie Arouet de Voltaire, French Author/Philosopher (1694–1778)

40.35 Since the foundation of that town [Nantucket] no epidemical distempers have appeared, which at times cause such depopulations in other countries; many of them are extremely well acquainted with the Indian methods of curing simple diseases and practise them with success. You will hardly find anywhere a community composed of the same number of individuals possessing such uninterrupted health.

Hector St. John de Crevecoeur, American Farmer/French Consul to New York (1735–1813)

40.36 This is much more prevailing among the women than the men, few of the latter having caught the contagion, though the sheriff, whom I may call the first person in the island [Nantucket], who is an eminent physician beside and who I had the pleasure of being well acquainted with, has for many years submitted to this custom [taking opium]. He takes three grains of it every day after breakfast, without the effects of which, he often told me, he was not able to transact any business.

Hector St. John de Crevecoeur, American Farmer/French Consul to New York (1735–1813)

40.37 Even the taking of medicine serves to make time go on with less heaviness. I have a sort of genius for physic and always had great entertainment in observing the changes of the human body and the effects produced by diet, labor, rest, and physical operations.

James Boswell, Scottish Lawyer/Author (1740–1795)

40.38 I have sometimes experienced from nitrous oxide, sensations similar to no others, and they have consequently been indescribable. This has been likewise often the case with other persons. Of two paralytic patients who were asked what they felt after breathing nitrous oxide, the first answered, "I do not know how, but very queer." The second said, "I felt like the sound of a harp." [On his pioneer experiments with the anesthetic nitrous oxide, or laughing gas.]

Sir Humphrey Davy, English Chemist (1778–1829)

40.39 As the heat of the weather increased, we heard of much sickness around us. The city is full of physicians, and they were all to be seen driving about in their cabs at a very alarming rate. One of these gentlemen told us, that when a medical man intended settling in a new situation, he always, if he knew his business, walked through the streets at night, before he decided. If he saw the dismal twinkle of the watch-light from many windows, he might be sure that disease was busy, and that the "location" might suit him well. Judging by this criterion, Cincinnati was far from healthy.

Frances Trollope, English Writer (1780–1863)

40.40 It is by no means easy to enjoy the beauties of American scenery in the west, even when you are in a neighborhood that affords much to admire; at least in doing so, you run considerable risk of injuring your health. Nothing is considered more dangerous than exposure to midday heat except exposure to evening damp; and the twilight is so short, that if you set out on an expedition when the fervid heat subsides, you can hardly get half a mile before "sun-down," as they call it, warns you that you must run or drive home again, as fast as possible, for fear you should get a "chill."

Frances Trollope, English Writer (1780–1863)

40.41 To Plato, the science of medicine appeared to be of very disputable advantage. He did not indeed object to quick cures for acute disorders, or for injuries produced by accidents. But the art which resists the slow sap of a chronic disease, which repairs frames enervated by lust, swollen by gluttony, or inflamed by wine, which encourages sensuality by mitigating the natural punishment of the sensualist, and prolongs existence when the intellect has ceased to retain its entire energy, had no share of his esteem. A life protracted by medical skill he pronounced to be a long death. The exercise of the art of medicine ought, he said, to be tolerated so far as that art may serve to cure the occasional distempers of men whose constitutions are good. As to those who have bad constitutions, let them die; and the sooner the better.

Thomas Babington Macaulay, English Historian/Author/Statesman (1800–1859)

40.42 Refuse to be ill. Never tell people you are ill, never own it to yourself. Illness is one of those things which a man should resist on principle.

Edward Bulwer-Lytton, English Politician/Author (1803–1873)

40.43 If I were a physician I would try my patients thus. I would wheel them to a window and let Nature feel their pulse. It will soon appear if their sensuous existence is sound. The sounds are but the throbbing of some pulse in me.

Henry David Thoreau, American Writer/Naturalist (1817–1862)

40.44 Let Nature do your bottling and your pickling and preserving. For all Nature is doing her best each moment to make us well. She exists for no other end. Do not resist her. With the least inclination to be well, we should not be sick. Men have discovered—or think they have discovered—the salutariness of a few wild things only, and not of all nature. Why, "nature" is but another name for health, and the seasons are but different states of health. Some men think that they are not well in spring, or summer, or autumn, or winter; it is only because they are not *well in* them.

Henry David Thoreau, American Writer/Naturalist (1817–1862)

40.45
Unnecessary noise is the most cruel absence of care which can be inflicted either on sick or well.

Florence Nightingale, English Nurse/Philanthropist (1820–1910)

40.46 Disease is an experience of so-called mortal mind. It is fear made manifest on the body.

Mary Baker Eddy, American Theologian/Christian Scientist (1821–1910)

40.47 It is a terrifying thought that life is at the mercy of the multiplication of these minute bodies, it is a consoling hope that Science will not always remain powerless before such enemies, since for example at the very beginning of the study we find that simple exposure to air is sufficient at times to destroy them.

Louis Pasteur, French Chemist/Microbiologist (1822–1895)

40.48 Care more for the individual patient than for the special features of the disease.

Sir William Osler, Canadian Physician/Anatomist (1849–1919)

40.49 When I worked on the polio vaccine, I had a theory. Experiments were done to determine what might or might not occur. I guided each one by imagining myself in the phenomenon in which I was interested. The intuitive realm is constantly active—the realm of imagination guides my thinking.

Jonas Salk, American Medical Researcher/Microbiologist (1915–)

40.50 Drug use, like war, is not a medical problem but a social, political, and economic one.

Dr. Henry L. Lennard, American Physician/Author (1971)

40.51 If more of our resources were invested in preventing sickness and accidents, fewer would have to be spent on costly cures. . . . In short, we should build a true "health" system—and not a "sickness" system alone.

Richard Milhous Nixon, American President (1971)

40.52 Their [city children's] environment is their major health problem. Rats, vermin, lead paint. They live in crowded tenements, many have no heat all winter. Most are on welfare and don't get the proper nutrition their little bodies need. They have health problems other children never even dream of.

Dr. Evelyn Schmidt, American Physician/Administrator (1971)

40.53 Medicine seems to be sharpening its tools to do battle with death as though death were just one more disease.

Dr. Leon R. Kass, American Science Society Administrator (1971)

40.54 We can cure most of the chronic cases given up as hopeless by the Western medicine men.

Dr. S.A. Subhan, Indian Government Hospital Administrator (1971)

40.55 10 Golden Rules for Good Health:

1. Have a checkup every year.
2. Be a non-smoker.
3. Drink in moderation.
4. Count each calorie.
5. Watch your cholesterol.
6. Learn nutritional values.
7. Find time for leisure and vacations.
8. Adjust to life's daily pressures.
9. Develop an exercise program.
10. Understand your physical assets and limitations.

The American Health Foundation (1982)

40.56 Here was a force [malpractice charges] taking away the physician's humanity—their right to make a mistake, and their right to intuition. A complex system of defensive medicine was developing, further driving up costs by high insurance premiums and because physicians would over-order tests to protect themselves. Diseases began to have standard treatments, and even if they weren't successful, it was better to do those than try something new, because the threat of malpractice charges brought fear. How can the doctor give meaningful medicine if every interaction has this element of fear?

Patch Adams, American Physician (1986)

40.57 There are many similarities between tumor cells and cells of the fetal-placental unit.

Paul Volberding, American Researcher (1986)

40.58 These estimates would make radon the second-greatest cause, after smoking, of America's total lung cancer death rate.

Environmental Protection Agency (1986)

40.59 Physicians are becoming more attuned to the necessity for individualizing drug doses. That's part of the art of being a good physician. . . . [otherwise] they'll kill their patients.

Elliott Vesell, American Pharmacogeneticist (1986)

40.60 As we begin to understand more of the determinants of susceptibility—which probably won't be reasonably complete for another two lifetimes—we may be able to develop a type of genetic printout when we're born, [to] indicate those kinds of environmental insults to avoid because of our own biochemistry.

Edward Calabrese, American Toxicologist (1986)

40.61 As medicine becomes more commercialized, what was defined as unscrupulous behavior is redefined as normal business. Where once we took pride in institutions that served the public and served it well, we now measure status by the bottom line.

Keith Schnieder, American Journalist (1986)

40.62 What are survivors like? If there is profile we can share with others . . . Survivors, in general, are nonconformists, "bad patients." In one study of the immune system . . . there was almost a 100 percent correlation between the head nurse's opinion of the patient and his immune-system function: a bad opinion meant a good immune system.

Bernard Siegal, American Surgeon (1986)

40.63 Each person's physical health may be affected by the health status of his or her neighbor. . . . The health of others affects our psychological well-being. . . . Decent people—and we are a decent people—are offended by unnecessary pain and suffering, that is by pain and suffering for which there is a treatment and for which some (who are affluent) are treated.

Rashi Fein, American Medical Writer (1986)

40.64 Jenner's smallpox vaccine remains the most effective vaccine ever produced. It actually eradicated a virus. Smallpox is the only disease that has been completely wiped off the face of the earth.

Bernard Moss, American Virologist (1986)

40.65 Eventually the process of aging, which is unlikely to be simple, should be understandable. Hopefully some of its processes can be slowed down or avoided. In fact, in the next century, we shall have to tackle the question of the preferred form of death.

Francis Crick, British Biophysicist/Geneticist (1986)

41. Meteorology

No one has a sorrier lot than the weatherman. He is ignored when he is right, but execrated when he is wrong.

ISAAC ASIMOV

41.1 While I was staying in Carignan, I twice during the night saw portents in the sky. These were rays of light towards the north, shining so brightly that I had never seen anything like them before: the clouds were blood-red on both sides, to the east and to the west. On a third night these rays appeared again, at about seven or eight o'clock. As I gazed in wonder at them, others like them began to shine from all four quarters of the earth, so that as I watched they filled the entire sky. A cloud gleamed bright in the middle of the heavens, and these rays were all focused on it, as if it were a pavilion the colored stripes of which were broad at the bottom but became narrower as they rose, meeting in a hood at the top. In between the rays of light there were other clouds flashing vividly as if they were being struck by lightning. This extraordinary phenomenon filled me with foreboding, for it was clear that some disaster was about to be sent from heaven.

Gregory of Tours, Frankish Ecclesiastic/Historian (538–594)

41.2 Thereafter there were many sailors on the sea and on inland water who said that they had seen a great and extensive fire near the ground in the north-east which continuously increased in width as it mounted to the sky. And the heavens opened into four parts and fought against it as if determined to put it out, and the fire stopped rising upwards. They saw that fire at the first streak of dawn, and it lasted until full daylight: this happened on 7 December.

Anonymous (1121)

41.3 The rainbow is the repercussion or refraction of rays of the sun in a concave aqueous cloud.

Robert Grosseteste, English Bishop/Educator (1168–1253)

41.4 Prospero: Hast thou, spirit,
Performed, to point, the tempest that I bade thee?
Ariel: To every article.
I boarded the king's ship. Now on the beak,
Now in the waist, the deck, in every cabin,
I flamed amazement. Sometime I'd divide
And burn in many places; on the topmast,
The yards, and boresprit would I flame distinctly,
Then meet and join. Jove's lightnings, the precursors
O' th' dreadful thunderclaps, more momentary
And sight-outrunning were not. The fire and cracks
Of sulfurous roaring the most mighty Neptune
Seem to besiege, and make his bold waves tremble;
Yea, his dread trident shake.

William Shakespeare, English Dramatist/Poet (1564–1616)

41.5 These days at ten o'clock at night a most alarming wonder has manifested itself in the skies. The firmament was rent asunder and through this gap one could distinguish chariots and armies, though to do battle against each other. This awesome and unusual vision continued from ten at night till about two of the morning, and was witnessed with alarm and dismay by many honest and trustworthy people. The significance thereof is known but to God Almighty, Who may graciously prevent the shedding of innocent blood.

Anonymous, Business Agent for the Fugger Banking Family (1589)

41.6 The frost continuing more and more severe, the Thames before London was still planted with booths in formal streets . . . so that it see'd to be a bacchanalian triumph or carnival on the water, whilst it was a severe judgement on the land, the trees not only splitting as if lightning-struck, but men and cattle perishing in diverse places, and the very seas so lock'd up with ice, that no vessels could stir out or come in. London, by reason of the smoke, was so filled with the fuliginous steame of the sea-coale, that hardly could one see crosse the streets, and this filling the breast, so as one could hardly breath. Here was no water to be had from the pipes and engines, nor could the brewers and divers other tradesmen worke, and every moment was full of disastrous accidents.
[Describing the Great Frost, 1683–1684.]

John Evelyn, English Diarist (1620-1706)

41.7 It is the flash which appears, the thunderbolt will follow.

François-Marie Arouet de Voltaire, French Author/Philosopher (1694–1778)

41.8 Of all the departments of science no one seems to have been less advanced for the last hundred years than that of meteorology.
Thomas Jefferson, American President/Author (1743–1826)

41.9 Looking at the thunder machine which had been set up, I saw not the slightest indication of the presence of electricity. However, while they were putting the food on the table, I obtained extraordinary electric sparks from the wire. My wife and others approached from it, for the reason that I wished to have witnesses see the various colors of fire about which the departed Professor Richmann used to argue with me. Suddenly it thundered most violently at the exact time that I was holding my hand to the metal, and sparks crackled. All fled away from me, and my wife implored that I go away. Curiosity kept me there two or three minutes more, until they told me that the soup was getting cold. By that time the force of electricity greatly subsided. I had sat at table only a few minutes when the man servant of the departed Richmann suddenly opened the door, all in tears and out of breath from fear. I thought that some one had beaten him as he was on his way to me, but he said, with difficulty, that the professor had been injured by thunder. . . . Nonetheless, Mr. Richmann died a splendid death, fulfilling a duty of his profession.
Mikhail Vasilievich Lomonosov, Russian Chemist/Physicist/Poet (1711–1765)

41.10 One source of deep interest to us, in this new clime, was the frequent recurrence of thunder-storms. Those who have only listened to thunder in England, have but a faint idea of the language which the gods speak when they are angry. . . . Every thing seems colossal on this great continent; if it rains, if it blows, if it thunders, it is all done *fortissimo;* but I often felt terror yield to wonder and delight; so grand, so glorious were the scenes a storm exhibited.
Frances Trollope, English Writer (1780–1863)

41.11 Every dew-drop and rain-drop had a whole heaven within it.
Henry Wadsworth Longfellow, American Poet (1807–1882)

41.12 Every leaf and twig was this morning covered with a sparkling ice armor; even the grasses in exposed fields were hung with innumerable diamond pendants, which jingled merrily when brushed by the foot of the traveler. It was literally the wreck of jewels and the crash of gems.
Henry David Thoreau, American Writer/Naturalist (1817–1862)

41.13 A sky without clouds is a meadow without flowers, a sea without sails.
Henry David Thoreau, American Writer/Naturalist (1817–1862)

41.14 Once I dipt into the future far as human eye could see,
And I saw the Chief Forecaster, dead as any one can be—
Dead and damned and shut in Hades as a liar from his birth,
With a record of unreason seldom paralleled on earth.
While I looked he reared him solemnly, that incandescent youth,

From the coals that he's preferred to the advantages of truth.
He cast his eyes about him and above him; then he wrote
On a slab of thin asbestos what I venture here to quote—
For I read it in the rose-light of the everlasting glow:
"Cloudy; variable winds, with local showers; cooler; snow."

Ambrose Bierce, American Satirist (1842–1914?)

41.15 Louisville looked like a new Atlantis with from three-fourths to four-fifths of its total area almost eavedeep in dark brown water. Cincinnati, except in a few high spots, seemed planted in a chocolate sea. Near Evansville, Indiana, where the Ohio seldom broadens out more than a half mile or quarter mile, the waters had spread from fifteen to twenty miles beyond the river banks. The river seemed to cover a dead world, with barnroofs and treetops and farmhouses barely visible above the tide.
[A reporter's description of an aerial view of the twenty-five day flood of 1937—the most widespread flood that ever occurred in America.]

***The New York Times* (1937)**

41.16 The weatherman, unlike the pilot, need not guess. He has got a slide rule; he has got the laws of gases, Charles's Law, Boyle's Law, Buys Ballot's Law at his fingertips. He has studied thermodynamics, and he has got a new device that is the biggest thing in weather science since Torricelli invented the barometer—the radio sonar with which he can take soundings of the upper air.

Wolfgang Langewiesche, American Science Journalist/Aeronauticist (1942)

41.17 A rainfall of fish occurred on October 23, 1947, in Marksville, Louisiana, while I was conducting biological investigations for the Department of Wild Life and Fisheries. In the morning of that day, between seven and eight o'clock, fish ranging from two to nine inches in length fell on the streets and in yards, mystifying the citizens of that southern town. I was in the restaurant with my wife having breakfast, when the waitress informed us that fish were falling from the sky.

A.D. Bajkov, American Biologist (1949)

41.18 Steadily the tornado came on, the end gradually rising above the ground. I could have stood there only few seconds, but so impressed was I with what was going on that it seemed a long time. At last the great shaggy end of the funnel hung directly overhead. Everything was as still as death. There was a strong gassy odor, and it seemed that I could not breathe. There was a screaming, hissing sound coming directly from the end of the funnel. I looked up and to my astonishment I saw right up into the heart of the tornado. There was a circular opening in the center of the funnel, about fifty or one hundred feet in diameter, and extending straight upward for a distance of at least one half mile. . . . The walls of this opening were of rotating clouds and the whole was made brilliantly visible by constant flashes of lightning which zigzagged from side to side.

Will Keller, American Farmer (1957)

41.19 Methane is released by bogs, and some 45 million tons of the same gas, it has been calculated, are added to the atmosphere each year by the venting of intestinal gases by cattle and other large animals.

Isaac Asimov, American Biochemist/Author (1965)

41.20 Over the next 88 years, ozone depletion could result in 40 million cases of skin cancer in the United States, 800,000 of which would end in death. [From a draft of a study of the potential health effects of the thinning of the stratospheric ozone layer, by the Environmental Protection Agency.]

Environmental Protection Agency (1986)

41.21 By the end of the next century, the "greenhouse effect" may increase temperatures worldwide to levels that have not been reached for at least 100,000 years. And the effects on sea level and on agriculture and other human activities are likely to be so profound that we should be planning for them now.

John Gribbin, American Science Journalist (1986)

42. Microbiology

Everything about microscopic life is terribly upsetting. How can things so small be so important?

ISAAC ASIMOV

42.1 In the year 1657 I discovered very small living creatures in rain water.
Antonie van Leeuwenhoek, Dutch Scientist/Inventor (1632–1723)

42.2 There is scarce a single Humour in the Body of a Man, or of any other Animal, in which our Glasses do not discover Myriads of living Creatures.
Joseph Addison, English Essayist/Poet/Politician (1672–1719)

42.3 Upon viewing the *milt* or *semen Masculinum* of a living Codfish with a *Microscope*, such Numbers of *Animalcules* with long Tails were found therein, that at least ten thousand of them were supposed to exist in the quantity of a Grain of Sand.
Henry Baker, English Microbiologist/Poet (1698–1774)

42.4 The joy I felt at the prospect before me of being the instrument destined to take away from the world one of its greatest calamities [smallpox] . . . was so excessive that I sometimes found myself in a kind of reverie.
Edward Jenner, English Naturalist (1803–1872)

42.5 To demonstrate experimentally that a microscopic organism actually is the cause of a disease and the agent of contagion, I know no other way, in the present state of Science, than to subject the *microbe* to the method of cultivation out of the body.
Louis Pasteur, French Chemist/Microbiologist (1822–1895)

42.6 One day when the whole family had gone to the circus to see some extraordinary performing apes, I remained alone with my microscope, observing the life in the mobile cells of a transparent starfish larva, when a new thought suddenly flashed across my brain. It struck me that similar cells might serve in the defense of the organism against intruders. Feeling that there was in this something of surpassing interest, I felt so excited that I began striding up and down the room and even went to the seashore to collect my thoughts.
[On his discovery of phagocytes.]

Ilya Ilich Metchnikov, Russian Zoologist (1845–1916)

42.7 When I look back upon the past, I can only dispel the sadness which falls upon me by gazing into that happy future when the infection [puerperal fever] will be banished. But if it is not vouchsafed for me to look upon that happy time with my own eyes . . . the conviction that such a time must inevitably sooner or later arrive will cheer my dying hour.
[Regarding the opposition to his discovery that puerperal fever was transmitted to women by the "cadaveric material" on the hands of physicians coming direct from the post-mortem room to an examination. He discovered that washing the physician's hands in a solution of chlorinated lime prevented the transmission of this disease. His doctrine was opposed by many, and he died of the very disease he sought to cure in an insane asylum in 1865.]

Ignaz Semmelweis, Hungarian Physician (1847)

42.8 It is usually not recognized that for every injurious or parasitic microbe there are dozens of beneficial ones. Without the latter, there would be no bread to eat nor wine to drink, no fertile soils and no potable waters, no clothing and no sanitation. One can visualize no form of higher life without the existence of the microbes. They are the universal scavengers. They keep in constant circulation the chemical elements which are so essential to the continuation of plant and animal life.

Selman Abraham Waksman, American Microbiologist (1888–1973)

42.9 Microbiology is usually regarded as having no relevance to the feelings and aspirations of the man of flesh and bone. Yet, never in my professional life do I find myself far removed from the man of flesh and bone. It is not only because microbes are ubiquitous in our environment, and therefore must be studied for the sake of human welfare. More interesting, and far more important in the long run, is the fact that microbes exhibit profound resemblances to man. They resemble him in their physical makeup, in their properties, in their responses to various stimuli; they also display associations with other living things which have perplexing and illuminating analogies with human societies.

René J. Dubos, French/American Bacteriologist (1901–1982)

42.10 What do you get when you cross a tobacco plant with a firefly? Not a cigarette that lights itself, a team of scientists reported from the University of

California, San Diego, reported last week, but an otherwise normal plant that emits a faint but constant light.

Laura Mansnerus and Katherine Roberts, American Science Journalists (1986)

42.11 With NMR [nuclear magnetic resonance imaging microscope] we don't touch the cell at all. We can follow it through time, watch its development.

James Aguayo, American Biology Researcher (1986)

42.12 Bacteria invented fermentation, the wheel in the form of the proton rotary motor, sulphur breathing photosynthesis, and nitrogen fixation, long before our evolution. They are not only social beings, but behave as a sort of worldwide decentralized democracy.

Lynn Margulis and Dorion Sagan, American Biologist/Author and American Science Journalist (1986)

42.13 Viral nucleic acids can become integrated into the chromosomes of their host cells, perhaps to evolve into "genes" that sometimes help us but sometimes cause disease. It seems equally likely that during the course of evolution genetic material originally belonging to cells has "escaped" to evolve into the genomes of infectious viruses (and various subviral agents). This is the favored explanation for the origin of most viruses, and there is good evidence to suggest that it is continuing to this day.

Andrew Scott, English Science Journalist (1986)

42.14 As medical researchers have gained more experience with recombinant medicines, they have learned to anticipate side effects, and not to expect miracles. No longer do researchers expect to clone "cures." Instead, they are learning—one by one—the complex ways in which individual human proteins behave and how to use these proteins to make modest, but significant, inroads against disease.

Mary Murray, American Science Journalist (1987)

42.15 Adam
Had 'em.
[Microbes.]

Anonymous (1987)

43. Mineralogy

It is strange, but rocks, properly chosen

and polished, can be as beautiful as flowers,

and much more durable.

ISAAC ASIMOV

43.1 Gold is found in our own part of the world; not to mention the gold extracted from the earth in India by the ants, and in Scythia by the griffins. Among us it is procured in three different ways; the first of which is in the shape of dust, found in running streams. . . . A second mode of obtaining gold is by sinking shafts or seeking among the debris of mountains. . . . The third method of obtaining gold surpasses the labors of the giants even: by the aid of galleries driven to a long distance, mountains are excavated by the light of torches, the duration of which forms the set times for work, the workmen never seeing the light of day for many months together.

Pliny the Elder (Gaius Plinius Secundus), Roman Naturalist/Historian (23–79)

43.2 We say that, in very truth the productive cause is a mineralizing power which is active in forming stones. . . . This power, existing in the particular material of stones, has two instruments according to different natural conditions.

One of these is heat, which is active in drawing out moisture and digesting the material and bringing about its solidification into the form of stone, in Earth that has been acted upon by unctuous moisture . . . .

The other instrument is in watery moist material that has been acted upon by earthy dryness; and this [instrument] is cold, which . . . is active in expelling moisture.

Albertus Magnus, German Clergyman/Theologian/Scholar (1200?–1280)

43.3 Sarcophagus is a stone that devours dead bodies. . . . For this reason stone monuments are called *sarcophagi*.

Albertus Magnus, German Clergyman/Theologian/Scholar (1200?–1280)

43.4 And this is a miracle of nature in part known, namely, that iron follows the part of a magnet that touches it, and flies from the other part of the same magnet. And the iron turns itself after moving to the part of the heavens conformed to the part of the magnet which it touched.

Roger Bacon, English Philosopher/Clergyman (1220–1292)

43.5 If the Tincture of the Philosophers is to be used for transmutation, a pound of it must be projected on a thousand pounds of melted Sol [gold]. Then, at length, will a medicine have been prepared for transmuting the leprous moisture of the metals. This work is a wonderful one in the light of nature, namely, that by the Magistery, or the operation of the Spagyrist, a metal, which formerly existed, should perish, and another be produced. This fact has rendered that same Aristotle, with his ill-founded philosophy, fatuous.

Paracelsus (Theophrastus Bombastius), Swiss/German Chemist/Physician (1493–1541)

43.6 There are reported to be six species of metals, namely, gold, silver, iron, copper, tin, and lead. Actually there are more. Mercury is a metal although we differ on this point with the chemists. *Plumbum cinereum* (gray lead) which we call *bisemutum* was unknown to the older Greek writers. On the other hand, Ammonius writes correctly many metals are unknown to us, as well as many plants and animals.

Georgius Agricola, German Physician/Politician/Mineralogist (1494–1555)

43.7 Enhydros is a variety of geode. The name comes from the water it contains. It is always round, smooth, and very white but will sway back and forth when moved. Inside it is a liquid just as in an egg, as Pliny, our Albetus, and others believed, and it may even drip water. Liquid bitumen, sometimes with a pleasant odor, is found enclosed in rock just as in a vase.

Georgius Agricola, German Physician/Politician/Mineralogist (1494–1555)

43.8 Your Grace will no doubt have learnt from the weekly reports of one Marco Antonio Bragadini, called Mamugnano. . . . He is reported to be able to turn base metal into gold. . . . He literally throws gold about in shovelfuls. This is his recipe: he takes ten ounces of quicksilver, puts it into the fire, and mixes it with a drop of liquid, which he carries in an ampulla. Thus it promptly turns into good gold. He has no other wish but to be of good use to his country, the Republic. The day before yesterday he presented to the Secret Council of Ten two ampullas with this liquid, which have been tested in his absence. The first test was found to be successful and it is said to have resulted in six million ducats. I doubt not but that this will appear mighty strange to your Grace.

Anonymous, Business Agent for the Fugger Banking Family (1589)

43.9 Concerning the alchemist, Mamugnano, no one harbors doubts any longer about his daily experiments in changing quicksilver into gold. It was realized that his craft did not go beyond one pound of quicksilver. . . . Thus the belief is now held that his allegations to produce a number of millions have been a great fraud.

Anonymous, Business Agent for the Fugger Banking Family (1589)

43.10 I then spied a great stone, and sitting a while upon it, I fell to weigh in my thoughts that that stone was in a happier condition in some respects than either those sensitive creatures or vegetables I saw before, in regard that that stone, which propagates by assimilation, as the philosophers say, needed neither grass nor hay, or any aliment for restoration of nature, nor water to refresh its roots, or the heat of the sun to attract the moisture upwards to increase growth as the other did.

James Howell, Welsh Author (1594–1666)

43.11 July 11, 1656. Came home by Greenwich ferry, where I saw Sir J. Winter's project of charring sea-coal to burn out the sulphur and render it sweet [coke]. He did it by burning the coals in such earthen pots as the glassmen melt their metal, so firing them without consuming them, using a bar of iron in each crucible, or pot, which bar has a hook at one end, that so the coals being melted in a furnace with other crude sea-coals under them, may be drawn out of the pots sticking to the iron, whence they are beaten off in great half-exhausted cinders, which being rekindled make a clear pleasant chamber-fire deprived of their sulphur and arsenic malignity. What success it may have, time will discover.

John Evelyn, English Diarist (1620–1706)

43.12 *Hence* dusky *Iron* sleeps in dark abodes,
And ferny foliage nestles in the nodes;
Till with wide lungs the panting bellows blow,
And waked by fire the glittering torrents flow;
—Quick whirls the wheel, the ponderous hammer falls,
Loud anvils ring amid the trembling walls,
Strokes follow strokes, the sparkling ingot shines,
Flows the red slag, the lengthening bar refines;
Cold waves, immersed, the glowing mass congeal,
And turn to adamant the hissing Steel.

Erasmus Darwin, English Physician/Poet (1731–1802)

43.13 To learn . . . the ordinary arrangement of the different strata of minerals in the earth, to know from their habitual colocations and proximities, where we find one mineral; whether another, for which we are seeking, may be expected to be in its neighborhood, is useful.

Thomas Jefferson, American President/Author (1743–1826)

43.14 Ask why God made the GEM so small,
And why so huge the granite?
Because God meant, mankind should set
That higher value on it.

Robert Burns, Scottish Poet (1759–1796)

43.15 The essence of the simplest mineral phenomenon is as completely unknown to chemists and physicists today as is the essence of intellectual phenomenon to physiologists.

Claude Bernard, French Physiologist (1813–1878)

43.16 Aluminum was once a precious metal.

Jules Verne, French Author (1828–1905)

43.17 Diamonds have been objects of interest to all classes, but more especially to scientists and savants, to whom, even up to this present age, they are a mystery as to their origin or formation. Some attribute them to be of celestial origin, as aerolites, possessing electric light; others believe them of vegetable origin, since some are found with water cavities and also vegetable as well as animal matter embedded in them. Workers in them seem to have a more true and practical knowledge of them, and feel convinced that they are more of a volcanic origin.

***Scientific American* (1886)**

43.18 The Spaniards plundered Peru for its gold, which the Inca aristocracy had collected as we might collect stamps, with the touch of Midas. Gold for greed, gold for splendor, gold for adornment, gold for reverence, gold for power, sacrificial gold, life-giving gold, gold for tenderness, barbaric gold, voluptuous gold.

Jacob Bronowski, Polish/British Mathematician/Science Writer (1908–1974)

43.19 The wonder of crystals is the network of atoms that is constantly repeated.

Italo Calvino, Italian Author (1923–1985)

43.20 Nature's operations in laying down the world's mineral deposits have been proceeding over a span of probably two billion years. Now comes man digging into the storehouse.

C.C. Furnas, American Chemical and Aeronautical Engineer (1939)

43.21 What is the most common mineral on the earth's surface? Most people would quickly answer, "Quartz," but this is wrong: quartz is second. Water is a mineral, the most abundant species on the surface of our planet.

William B. Sanborn, American Rock Collector/Author (1976)

43.22 To the novice it seems incredible to dig out a hard, grayish, round rock, ring it with gentle taps of a rock hammer, and have it pop open, spilling dark black oil on your hands!
[Describing bitumen-filled geodes found near Niota, Illinois.]
William B. Sanborn, American Rock Collector/Author (1976)

43.23 Quartz crystals coated with proteins such as antibodies, enzymes, and other biologically active materials may be useful for detecting traces of pesticides and drugs in air.
Ivars Peterson, American Science Journalist (1986)

44.
The Moon

The Moon and its phases gave man his first calendar. Trying to match that calendar with the seasons helped give him mathematics. The usefulness of the calendar helped give rise to the thought of beneficent gods. And with all that the Moon is beautiful, too.

ISAAC ASIMOV

44.1 Splendor of the moon shines down on quiet night,
And night is quiet, quenching fumes and dust.
Moonbeams flood in at the doors,
While rounded flecks come filtered, through the cracks.
On her high tower they wound the yearning wife;
In western garden sport with men of talent;
Through curtained casement gleam from pearl-sewn threads;
Before the gate shine on the verdant moss;
But in the inmost chamber where the dawn has not yet come—
The limpid radiance—ah, how far away!
Shen Yüeh, Chinese Poet (441–512)

44.2 It is the very error of the moon;
She comes more nearer earth than she was wont,
And makes men mad.
William Shakespeare, English Dramatist/Poet (1564–1616)

44.3 The moon, the governess of floods.
William Shakespeare, English Dramatist/Poet (1564–1616)

44.4 I feel sure that the surface of the moon is not perfectly smooth, free from inequalities, and exactly spherical, as a large school of philosophers considers with regard to the moon and the other heavenly bodies, but that, on the contrary, it is full of inequalities, uneven, full of hollows and protuberances, just like the surface of the earth itself, which is varied everywhere by lofty mountains and deep valleys.
Galileo Galilei, Italian Astronomer/Physicist (1564–1642)

44.5 It is a most beautiful and delightful sight to behold the body of the moon.
Galileo Galilei, Italian Astronomer/Physicist (1564–1642)

44.6 He made an instrument to know
If the moon shine at full or no;
That would, as soon as e'er she shone straight,
Whether 'twere day or night demonstrate;
Tell what her d'ameter to an inch is,
And prove that she's not made of green cheese.
Samuel Butler, English Satirist (1612–1680)

44.7 Clouds come from time to time
and bring to men a chance to rest
from looking at the moon.
Matsuo Bashō, Japanese Poet (1644–1694)

44.8 It is the moon, I ken her horn
That's blinkin in the lift sae hie;
She shines sae bright to wyle us hame,
But by my sooth she'll wait a wee!
Robert Burns, Scottish Poet (1759–1796)

44.9 The cold chaste Moon, the Queen of Heaven's bright isles,
Who makes all beautiful on which she smiles!
That wandering shrine of soft, yet icy flame,
Which ever is transform'd yet still the same,
And warms, but not illumines.
Percy Bysshe Shelley, English Poet (1792–1822)

44.10 This shall be the test of innocence—if I can hear a taunt, and look out on this friendly moon, pacing the heavens in queenlike majesty, with the accustomed yearning.
Henry David Thoreau, American Writer/Naturalist (1817–1862)

44.11 The moon, as the only natural satellite of the earth, plays an important role in the exploration of outer space and should be used exclusively in the interests of peace and for the good of all mankind.
Andrei Andreyevich Gromyko, Russian Foreign Minister (1909–)

44.12 We waited in the quiet gardens under a full moon, and I had the chance to ask Armstrong whether, looking up at the moon from the Earth, he could really believe that he had trod the surface of the Earth's satellite. "It does seem a bit unreal now," he replied.
[In Paris, during an around-the-world goodwill tour after the moon landing.]
Max W. Kraus, American Foreign Services Officer (1970)

44.13 We have so many things to do on earth. Why go to the moon? It is a great waste and the people must pay for it.
Chou En-Lai, Chinese Politician (1971)

44.14 [In] preliminary conclusion, the material from all three seas—the Sea of Tranquility, the Ocean of Storms, and the Sea of Fertility—is surprisingly similar in its petrological, mineralogical, and chemical composition, though certain details are different.
Aleksandr P. Vinogradov, Russian Geochemist (1971)

44.15 It [the moon] looks like it's been molded out of plaster of Paris.
[Spoken while aboard the Apollo 14 spacecraft in lunar orbit.]
Commander Edgar D. Mitchell, Jr., American Astronaut (1971)

44.16 Hey, you're not gonna believe this. It [the moon] looks just like the map.
[Spoken while aboard the Apollo 14 spacecraft in lunar orbit.]
Major Stuart A. Roosa, American Astronaut (1971)

44.17 The eye of the mind leaps again into space, searching for the fabulous dot of the lunar vehicle, so much does its incredible reality fill us once more with amazement and admiration.
Pope Paul VI (1971)

44.18 It [a moon rock] looks like a dirty potato. I'm afraid they'll have to clean up the moon before I go there.
Martha Mitchell, Wife of John Mitchell, American Attorney General (1971)

44.19 The Earth has no business possessing such a Moon. It is too huge—over a quarter Earth's diameter and about 1/81 of its mass. No other planet in the Solar System has even nearly so large a satellite.
Isaac Asimov, American Biochemist/Author (1976)

44.20 New computer calculations, based on observations from the Voyager satellites and ground-based telescopes, show that just about every moon—except the earth's—has experienced millions of years [sic] chaotic tumbling.
A tumbling moon falls end over end, twists sideways, speeds up, slows down, all the time obeying Newton's completely deterministic laws of motion, yet defy [sic] prediction in a way that scientists used to consider impossible.
James Gleick, American Science Journalist (1987)

45.
Natural Law

There is very little flexibility in the behavior of the Universe. What it does once, it does again.

ISAAC ASIMOV

45.1 Nature produces those things which, being continually moved by a certain principle contained in themselves, arrive at a certain end.
Aristotle, Greek Philosopher (384 B.C.–322 B.C.)

45.2 For it is the nature of that which is the same and remains in the same state always to produce the same effects, so either there will always be generation or corruption.
Aristotle, Greek Philosopher (384 B.C.–322 B.C.)

45.3 For ourselves, we may take as a basic assumption, clear from a survey of particular cases, that natural things are some or all of them subject to change.
Aristotle, Greek Philosopher (384 B.C.–322 B.C.)

45.4 Everything which comes to be, comes to be out of, and everything which passes away passes away into, its opposite or something in between. And the things in between come out of the opposites—thus colors come out of pale and dark. So the things which come to be naturally all are or are out of opposites.
Aristotle, Greek Philosopher (384 B.C.–322 B.C.)

45.5 Nature is the principle of movement of its own accord and not by accident.
Anicius Manlius Severinus Boethius, Roman Philosopher/Statesman (480?–524?)

45.6 Can the cause be reached from knowledge of the effect with the same certainty as the effect can be shown to follow from its cause? Is it possible for one effect to have many causes? If one determinate cause cannot be reached from the effect, since there is no effect which has not some cause, it follows that an effect, when it has one cause, may have another, and so that there may be several causes of it.

Robert Grosseteste, English Bishop/Educator (1168–1253)

45.7 But we must here state that we should not see anything if there were a vacuum. But this would not be due to some nature hindering species; for species is a natural thing, and therefore needs a natural medium; but in a vacuum nature does not exist.

Roger Bacon, English Philosopher/Clergyman (1220–1292)

45.8 There are certain things which can exist but do not, and others which do exist. Those which can be are said to exist *in potency,* whereas those which are, are said to exist *in act.*

Saint Thomas Aquinas, Italian Theologian/Scholar (1225–1274)

45.9 It is undoubted truth that everything in this world has a period fixed for its duration.

Niccolò Machiavelli, Italian Statesman/Political Philosopher (1469–1527)

45.10 The law of nature is so unalterable that it cannot be changed by God Himself.

Hugo de Grotius, Dutch Jurist/Philosopher/Diplomat (1538–1645)

45.11 The stars of heaven are free because
In amplitude of liberty
Their joy is to obey the laws.

William Watson, English Poet (1559?–1603)

45.12 When everything moves at the same time, nothing moves in appearance.

Blaise Pascal, French Mathematician/Philosopher/Author (1623–1662)

45.13 A body at rest remains at rest and a body in motion remains in uniform motion in a straight line unless acted upon by an external force; the acceleration of a body is directly proportioned to the applied force and is the direction of the straight line in which the force acts; and for every force there is an equal and opposite force in reaction.

Isaac Newton, English Physicist/Mathematician (1642–1727)

45.14 Nature and Nature's laws lay hid in night.
God said, "Let Newton be," and all was light.

Alexander Pope, English Poet/Satirist (1688–1744)

45.15 The laws of nature are the rules according to which effects are produced; but there must be a lawgiver—a cause which operates according to these rules. . . . The laws of navigation never steered a ship, and the law of gravity never moved a planet.

Thomas Reid, Scottish Philosopher (1710–1796)

45.16 The first causes of motion are not in matter; it receives motion and communicates it, but it does not produce it. . . . A movement which is not produced by another can only come from an action spontaneous and voluntary; inanimate bodies only act by it, and there is no true action without will.

Jean-Jacques Rousseau, Swiss/French Philosopher/Author (1712–1778)

45.17 The present state of the system of nature is evidently a consequence of what is in the preceding moment, and if we conceive of an intelligence which at a given instant comprehends all the relations of the entities of this universe, it could state the respective positions, motions, and general effects of all these entities at any time in the past or future.

Pierre Simon Laplace, French Astronomer/Mathematician (1749–1827)

45.18 When Science from Creation's face
Enchantment's veil withdraws,
What lovely visions yield their place
To cold material laws!

Thomas Campbell, English Poet (1777–1844)

45.19 'Tis a short sight to limit our faith in laws to those of gravity, of chemistry, of botany, and so forth.

Ralph Waldo Emerson, American Essayist/Philosopher/Poet (1803–1882)

45.20 The chessboard is the world, the pieces are the phenomena of the universe, the rules of the game are what we call the laws of Nature. The player on the other side is hidden from us, we know that his play is always fair, just, and patient. But also we know, to our cost, that he never overlooks a mistake, or makes the smallest allowance for ignorance.

Thomas Henry Huxley, English Biologist/Evolutionist (1825–1895)

45.21 Around the ancient track marched, rank on rank,
The army of unalterable law.

George Meredith, English Novelist/Poet (1828–1909)

45.22 People make the mistake of talking about "natural laws." There are no natural laws. There are only temporary habits of nature.

Alfred North Whitehead, English Mathematician/Philosopher (1861–1947)

45.23 It helps to remind oneself every so often that over everything that exists there are laws which never fail to operate, which come rushing, rather, to manifest and prove themselves upon every stone and upon every feather we let fall. So all erring consists simply in the failure to recognize the natural laws to which we are subject in the given instance, and every solution begins with our alertness and concentration, which gently draw us into the chain of events and restore to our will its balancing counterweights.

Rainer Maria Rilke, Austrian Poet (1875–1926)

45.24 There is no logical way to the discovery of these elemental laws. There is only the way of intuition, which is helped by a feeling for the order lying behind the appearance.

Albert Einstein, German/American Physicist (1879–1955)

45.25 There is no field of experience which cannot, in principle, be brought under some form of scientific law, and no type of speculative knowledge about the world which it is, in principle, beyond the power of science to give.

A.J. Ayer, British Philosopher (1910–)

45.26 Everything in space obeys the laws of physics. If you know these laws, and obey them, space will treat you kindly. And don't tell me man doesn't belong out there. Man belongs wherever he wants to go—and he'll do plenty well when he gets there.

Wernher von Braun, German/American Rocket Engineer (1912–1977)

46. Nature

There seems to be a feeling that anything that is natural must be good. Strychnine is natural.

ISAAC ASIMOV

46.1 Omnia quae secundam Naturam fiunt sunt habenda in bonis. [The works of Nature must all be accounted good.]
Marcus Tullius Cicero, Roman Orator/Statesman (106 B.C.–43 B.C.)

46.2 There are no reptiles, and no snake can exist there [Ireland]; for although often brought over from Britain, as soon as the ship nears land, they breathe the scent of the air, and die.
Bede, English Monk/Scholar (673?–735)

46.3 He who wishes to see how much nature and the heavens can do among us should come to gaze upon her.
Francesco Petrarch, Italian Poet/Scholar (1304–1374)

46.4 Fallen Petals rise
back to the branch—I watch:
Oh . . . butterflies!
Moritake, Japanese Poet/Shinto Priest (1452–1540)

46.5 Nature is often hidden, sometimes overcome, seldom extinguished.
Francis Bacon, English Philosopher/Essayist/Statesman (1561–1626)

46.6 We cannot command nature except by obeying her.
Francis Bacon, English Philosopher/Essayist/Statesman (1561–1626)

46.7 All this visible universe is only an imperceptible point in the vast bosom of nature. The mind of man cannot grasp it. It is in vain that we try to stretch our conceptions beyond all imaginable space; we bring before the mind's eye merely atoms in comparison with the reality of things. It is an infinite sphere, of which the center is everywhere, the circumference nowhere.

Blaise Pascal, French Mathematician/Philosopher/Author (1623–1662)

46.8 There is not so contemptible a plant or animal that does not confound the most enlarged understanding.

John Locke, English Philosopher (1632–1704)

46.9 Nature always springs to the surface and manages to show what she is. It is vain to stop or try to drive her back. She breaks through every obstacle, pushes forward, and at last makes for herself a way.

Nicolas Boileau-Despréaux, French Poet/Critic (1636–1711)

46.10 We, too, are silly enough to believe that all nature is intended for our benefit.

Bernard Le Bovier Sieur de Fontenelle, French Philosopher (1657–1757)

46.11 My bees, above any other tenants of my farm, attract my attention and respect; I am astonished to see that nothing exists but what has its enemy; one species pursues and lives upon the other: unfortunately, our king-birds are the destroyers of those industrious insects, but on the other hand, these birds preserve our fields from the depredation of crows, which they pursue on the wing with great vigilance and astonishing dexterity.

Thus divided by two interested motives, I have long resisted the desire I had to kill them until last year, when I thought they increased too much. . . . I killed him [a king-bird] and immediately opened his craw, from which I took 171 bees; I laid them all on a blanket in the sun, and to my great surprise, 54 returned to life, licked themselves clean, and joyfully went back to the hive, where they probably informed their companions of such an adventure and escape as I believe had never happened before to American bees!

Hector St. John de Crevecoeur, American Farmer/French Consul to New York (1735–1813)

46.12 I'd forgive Nature all the rest if she would rid us of these cunning, devouring thieves which no art can subdue. When the floods rise on our low grounds, the mice quit their burrows and come to our stacks of grain or to our heaps of turnips, which are buried under the earth out of the reach of the frost. There secured from danger, they find a habitation replenished with all they want. I must not, however, be murmuring and ungrateful. If Nature has formed mice, she has created also the fox and the owl. They both prey on these. Were it not for their kind assistance, [the mice] would drive us out of our farms.

Thus one species of evil is balanced by another; thus the fury of one element is repressed by the power of the other. In the midst of this great, this astonishing equipment, Man struggles and lives.

Hector St. John de Crevecoeur, American Farmer/French Consul to New York (1735–1813)

46.13 There is not a sprig of grass that shoots uninteresting to me.
Thomas Jefferson, American President/Author (1743–1826)

46.14 Nature seems to operate always according to an original general plan, from which she departs with regret and whose traces we come across everywhere.
Vic D'Azyr, French Anatomist/Physician (1748–1794)

46.15 Without my attempts in natural science, I should never have learned to know mankind such as it is. In nothing else can we so closely approach pure contemplation and thought, so closely observe the errors of the senses and of understanding.
Johann Wolfgang von Goethe, German Poet/Dramatist/Novelist (1749–1832)

46.16 Nature is beneficent. I praise her and all her works. She is silent and wise. She is cunning, but for good ends. She has brought me here and will also lead me away. She may scold me, but she will not hate her work. I trust her.
Johann Wolfgang von Goethe, German Poet/Dramatist/Novelist (1749–1832)

46.17 Come forth unto the light of things,
Let Nature be your teacher.
William Wordsworth, English Poet (1770–1850)

46.18 A large party, who had crossed from the American side, wound up the steep ascent from the place where the boat had left them; in doing so their backs were turned to the cataracts, and as they approached the summit our party was the principal object before them. They all stood perfectly still to look at us. . . . Then they advanced in a body, and one or two of them began to examine (wrong side upwards) the work of the sketcher, in doing which they stood precisely between him and his object. . . . Some among them next began to question us as to how long we had been at the falls; . . . In return we learnt that they were just arrived; yet not one of them (they were eight) ever turned the head, even for a moment, to look at the most stupendous spectacle [Niagara Falls] that nature has come to show.
Frances Trollope, English Writer (1780–1863)

46.19 Nature has no more concern or praise for human souls than for ants.
Giacomo Leopardi, Italian Poet (1798–1837)

46.20 The study of Nature is intercourse with the highest Mind. You should never trifle with Nature.
Jean Louis Agassiz, Swiss/American Naturalist/Geologist (1807–1873)

46.21 Go on, fair science: soon to thee
Shall nature yield her idle boast:

Her vulgar fingers formed a tree,
But thou hast trained it to a post.
Oliver Wendell Holmes, American Physician/Poet/Author (1809–1894)

46.22 Nature will tell you a direct lie if she can.
Charles Robert Darwin, English Naturalist/Evolutionist (1809–1882)

46.23 How indispensable to a correct study of Nature is a perception of her true meaning. The fact will one day flower out into a truth. The season will mature and fructify what the understanding had cultivated. Mere accumulators of facts—collectors of materials for the master-workmen—are like those plants growing in dark forests, which "put forth only leaves instead of blossoms."
Henry David Thoreau, American Writer/Naturalist (1817–1862)

46.24 Nature, with equal mind,
Sees all her sons at play,
Sees man control the wind,
The wind sweep man away.
Matthew Arnold, English Poet/Critic (1822–1888)

46.25 The investigation of nature is an infinite pasture-ground where all may graze, and where the more bite, the longer the grass grows, the sweeter is its flavor, and the more it nourishes.
Thomas Henry Huxley, English Biologist/Evolutionist (1825–1895)

46.26 In nature there are neither rewards nor punishments—there are consequences.
Robert Green Ingersoll, American Lawyer (1833–1899)

46.27 That man can interrogate as well as observe nature was a lesson slowly learned in his evolution.
Sir William Osler, Canadian Physician/Anatomist (1849–1919)

46.28 Let the mind rise from victory to victory over surrounding nature, let it but conquer for human life and activity not only the surface of the earth but also all that lies between the depth of the sea and the outer limits of the atmosphere; let it command for its service prodigious energy to flow from one part of the universe to the other, let it annihilate space for the transference of its thoughts.
Ivan Petrovich Pavlov, Russian Physiologist (1849–1936)

46.29 Men have brought their powers of subduing the forces of nature to such a pitch that by using them they could now very easily exterminate one another to the last man.
Sigmund Freud, Austrian Psychiatrist/Psychoanalyst (1856–1939)

46.30 In all the short lifetime of Danish summer there is no richer or more luscious moment than that wherein the lime trees flower. The heavenly scent goes to the head and to the heart; it seems to unite the fields of Denmark with those of Elysium; it contains both hay, honey, and holy incense, and is half fairy-land and half apothecary's locker.

Isak Dinesen (Baroness Karen Blixen), Danish Author (1885–1962)

46.31 The whole of Nature is a mighty struggle between strength and weakness.

Adolf Hitler, German Dictator (1889–1945)

46.32 Natural science is one of man's weapons in his fight for freedom. . . . For the purpose of attaining freedom in the world of nature, man must use natural science to understand, conquer, and change nature and thus attain freedom from nature.

Mao Tse Tung, Chinese Political Leader (1893–1976)

46.33 By putting nature to the most embarrassing tests he can devise, the scientist makes the most of his lay flair for evidence; and at the same time he amplifies the flair itself, affixing an artificial proboscis of punch cards and quadrille paper.

Willard van Orman Quine, American Philosopher (1908–)

46.34 I remember discussions with Bohr which went through many hours till very late at night and ended almost in despair; and when at the end of the discussion I went alone for a walk in the neighboring park I repeated to myself again and again the question: Can nature possibly be as absurd as it seemed to us in these atomic experiments?

Werner Karl Heisenberg, German Physicist (1958)

46.35 From beans to bacteria to the scavenger-hunter bond between larval crabs and jellyfish, symbiotic relationships show that peaceful coexistence is part of the very foundation of nature.

Robert Masello, American Author (1986)

47. Nuclear Energy

It was not until 1901 that humanity knew that nuclear energy existed. It is understandable now—but useless—to wish that we still lived in the ignorance of 1900.

ISAAC ASIMOV

47.1 To the village square, we must bring the facts about nuclear energy. And from here must come America's voice.
Albert Einstein, German/American Physicist (1879–1955)

47.2 All in all, the total amount of power conceivably available from the uranium and thorium supplies of the earth is about twenty times that available from the coal and oil we have left.
Isaac Asimov, American Biochemist/Author (1920–)

47.3 Contrary to widespread belief, nuclear power is no longer a cheap energy source. In fact, when the still unknown costs of radioactive waste and spent nuclear fuel management, decommissioning, and perpetual care are finally included on the rate base, *nuclear power may prove to be much more expensive* than conventional energy alternatives such as coal.
Congressional Report, United States Congressional Committee (1978)

47.4 Unprecedented opportunities for new scientific experience . . . immediate ground-floor opportunities for dedicated scientists and engineering personnel who want to be in the forefront of emerging technologies.

[Advertisement placed in classified section of *The New York Times* to recruit workers for the post-accident cleanup at Three Mile Island.]

Metropolitan Edison, Energy Company Advertisement (1979)

47.5 I found several robins and starlings laying around on top of the hay in my barn. They flew in there and died. Up at my brother's, the robins. flew in the porch and fell dead in a peach basket. It killed our pheasants. . . . It killed the quail too, no quail last summer. I didn't see a snake on the whole farm last summer. It killed the hoptoads.

Charles Conley, American Farmer—Three Mile Island area (1980)

47.6 The radiation just hung right there on the first day of the accident. There was a temperature inversion that morning. Then after the air began to move and the radioactivity began to move with it, they commenced monitoring. So they didn't measure the real heavy burst that was the first release. We don't know what we got. The equipment that they had here down along the creek, what they call a "sniffer," had been through two floods. So that thing was completely out of shape—the gate was hanging open with the lock off the door. That was supposed to have been the detector on this side of the river. It wasn't running at all. It was completely shot. . . .

I'd be quite curious to know what degree of radioactivity I was exposed to. It was a close call. I did taste the metal. They claim that's an indication of a high release. And I understand that there were people over in Middletown who had that metallic taste, too.

Bill Whittock, American Retired Civil Engineer (1980)

47.7 In a number of ways the accident is still in progress. First, of course, is the problem of releasing krypton gas into the atmosphere. . . . The next way the accident is still in progress pertains to the large quantities of radioactive water that remain on site. . . . The second body of water, which concerns us perhaps the most, is the highly contaminated seven hundred thousand gallons of water sitting in the basement on the containment building at TMI. Now, that concrete was not constructed as a swimming pool. Yet, it's been filled with hot water for over a year. There are concerns about seals, pipes, and potential leakage. If it leaks, it's the Susquehanna River and Chesapeake Bay that will be contaminated.

Dr. Judith Johnsrud, American Geographer/Environmentalist (1980)

47.8 The problem of the nuclear-power industry is that we have had too few accidents. . . . It's expensive, but that's how you gain experience.

Sigvard Eklund, General Director Atomic Energy Commission (1980)

47.9 That a shutdown of existing reactors would be catastrophic I believe is self-evident. It is not only the energy that we would lose, it is the $100 billion investment whose write-off would cause a violent shock to our financial institutions.

Alvin M. Weinberg, American Scientist (1980)

47.10 The point is that over and over again, the government and the industry have either lied to or misled the press and the public about nuclear energy. And the press and the public don't trust the industry. They don't trust the government. Whether that trust can be rebuilt is a very, very, tough question for our society.

David Burnham, American Environmentalist (1980)

47.11 I don't think the utility was covering up any information, but it just was difficult to piece together because there was so much information—so much information that it was very difficult to put it together into a comprehensive and logical story, not only to describe what had happened and the extent of the accident but to tell what might happen in the future.

Richard Vollmer, American Government Official (1980)

47.12 The most fundamental lesson of TMI, one that must be continually emphasized, is that accidents can happen.

Congressional Subcommittee on Energy Research and Production (1980)

47.13 The major recommendations of the report sidestep the issue of nuclear safety and stress the importance of addressing "the lack of understanding of nuclear power by the public.". . . In other words, according to the report, it is an uninformed public, rather than the accident at Three Mile Island, that is to be blamed for the present widespread concern about the competence of the nuclear industry and the adequacy of nuclear safety regulation.

Moreover, we would argue that the effort that has been made in this document to minimize the very serious safety problems that are implicit in nuclear power generation, and were dramatically highlighted by the accident at Three Mile Island, does little to restore the credibility of the nuclear option, or inspire public confidence in the objectivity of Congressional oversight.

Howard Wolpe and Richard Ottinger, American Congressmen (1980)

47.14 Human civilization is but a few thousand years long. Imagine having the audacity to think that we can devise a program to store lethal radioactive materials for a period of time that is longer than all of human culture to date.

Jeremy Rifkin, American Writer/Environmental Activist (1980)

47.15 We have found in our present study nothing inherent in reactors or in safeguard systems as they now have been developed which guarantees either that major reactor accidents will not occur or accidents occur and the protective systems will not fail. Should such accidents occur and the protective systems fail, very large damages could result.

Government Study for Brookhaven National Laboratory, Long Island (1982)

47.16 No existing [nuclear power] plant uses foolproof technology, and the consequences of a meltdown, no matter how remote the probability of its occurrence, are so serious that any benefits of nuclear power can *never* outweigh the risks.

Gerry Waneck, American Environmental Activist (1986)

47.17 Working at the [Chernobyl] station is safer than driving a car. [Spoken two months before the accident.]

Pyotr Bondarenko, Soviet Safety Official (1986)

47.18 Several highly improbable and therefore unforeseen failures. [Stated cause of accident at Chernobyl Nuclear Power Station, USSR, Saturday, April 26, 1986.]

Soviet Officials, Press Conference Release (1986)

47.19 Suddenly there was an enormous explosion, like a violent volcano. The nuclear reactions had led to overheating in the underground burial grounds. The explosion poured radioactive dust and materials high up into the sky. It was just the wrong weather for such a tragedy. Strong winds blew the radioactive clouds hundreds of miles away.

It was difficult to gauge the extent of the disaster immediately, and no evacuation plan was put into operation right away. Many villages and towns were only ordered to evacuate when the symptoms of radiation sickness were already quite apparent. Tens of thousands of people were affected, hundreds dying, though the real figures have never been made public. The large area, where the accident happened, is still considered dangerous and is closed to the public.

Zhores Medvedev, Russian Biochemist (1986)

47.20 I'm not suggesting that we're all about to kill ourselves with nuclear waste. I'm just saying we've created something that requires the highest levels of human quality perpetually, and I have little confidence that we are going to be up to that challenge.

Amory Lovins, American Physicist (1986)

47.21 The important lesson from the Chernobyl disaster is that the best brains in the world are needed urgently to work on developing safe nuclear energy, and that the warnings of these scientists must be heeded even when they say things that their governments and the nuclear power industry do not like.

Harry J. Lipkin, American Physicist (1986)

47.22 Neither the Nuclear Regulatory Commission nor the national laboratories have the skills to contain core materials that have melted through the bottom of a reactor's concrete base mat, preventing them from contaminating groundwater and migrating through soil. Unless the needed expertise is developed . . . officials might consider declaring such a site a national monument and keeping people away from it—perhaps for hundreds of years.
[From a study done while Niemczyk was working at Sandia National Laboratories in Albuquerque, NM.]

Susan Niemczyk, American Physicist (1986)

48. Nuclear Weapons

The mere existence of nuclear weapons by the thousands is an incontrovertible sign of human insanity.

ISAAC ASIMOV

48.1 I am very unhappy to conclude that the hydrogen bomb should be developed and built.
Harold Clayton Urey, American Physicist (1893–1981)

48.2 I am become Death, the destroyer of worlds.
J. Robert Oppenheimer, American Physicist (1945)

48.3 The effects could well be called unprecedented, magnificent, beautiful, stupendous, and terrifying. No man-made phenomenon of such tremendous power had ever occurred before. The lighting effects beggared description. The whole country was lighted by a searing light with the intensity many times that of the midday sun. It was golden, purple, violet, gray, and blue. It lighted every peak, crevasse, and ridge of the nearby mountain range with a clarity and beauty that cannot be described but must be seen to be imagined.
Brigadier General Thomas Farrell, American Soldier (1945)

48.4 There was a mushroom, of course, and under it the city seemed to just be a black, billowing layer of boiling tar.
[Describing Hiroshima after impact of atomic bomb dropped by the *Enola Gay* with Tibbetts as commander.]
Colonel Paul Tibbetts, Jr., American Soldier (1945)

48.5 One would have to have been brought up in the "spirit of militarism" to understand the difference between Hiroshima and Nagasaki on the one hand,

and Auschwitz and Belsen on the other. The usual reasoning is the following: the former case is one of warfare, the latter of cold-blooded slaughter. But the plain truth is that the people involved are in both instances nonparticipants, defenseless old people, women, and children, whose annihilation is supposed to achieve some political or military objective. . . . I am certain that the human race is doomed, unless its instinctive detestation of atrocities gains the upper hand over the artificially constructed judgment of reason.

Max Born, German/British Physicist (1953)

48.6 We will not act prematurely or unnecessarily risk the costs of world-wide nuclear war in which even the fruits of victory would be ashes in our mouth. But neither will we shrink from that risk at any time it must be faced. [From a television address during the Cuban missile crisis, on October 22.]

John F. Kennedy, American President (1962)

48.7 Certainly, speaking for the United States of America, I pledge that, as we sign this treaty in an era of negotiation, we consider it only one step toward a greater goal: the control of nuclear weapons on earth and the reduction of the danger that hangs over all nations as long as those weapons are not controlled.

Richard Milhous Nixon, American President (1971)

48.8 We are engaged in negotiations with the U.S.A. on a limitation of strategic armaments. Their favorable outcome would make it possible to avoid another round in the missile arms race and to release considerable resources for constructive purposes. We are seeking to have the negotiations produce positive results.

Leonid I. Brezhnev, Russian Communist Party Leader (1971)

48.9 But of all environments, that produced by man's complex technology is perhaps the most unstable and rickety. In its present form, our society is not two centuries old, and a few nuclear bombs will do it in.

To be sure, evolution works over long periods of time and two centuries is far from sufficient to breed Homo technikos. . . .

The destruction of our technological society in a fit of nuclear peevishness would become disastrous even if there were many millions of immediate survivors.

The environment toward which they were fitted would be gone, and Darwin's demon would wipe them out remorselessly and without a backward glance.

Isaac Asimov, American Biochemist/Author (1976)

48.10 There was a loudspeaker that reported on the time left before the blast: "T-minus ten minutes"—something like that. The last few seconds were counted off one by one. We had all turned away. At zero there was the flash. I counted and then turned around. The first thing I saw was a yellow-orange fireball that kept getting larger. As it grew, it turned more orange and then red. A mushroom-shaped cloud of glowing magenta began to rise over the desert where the explosion had been. My first thought was, "My God, that is beautiful!"

Jeremy Bernstein, American Physicist/Author (1982)

48.11 Today only the Soviet Union is prepared not only to wage war in space, but also to support Earth-based conflict using space resources.

Nicholas Johnson, American Aeronautics Engineer (1982)

48.12 The propaganda is being disseminated that the United States, supposedly, is developing space weaponry programs only to hear that similar projects had been launched in the USSR. All this is a premeditated lie, a propagandistic myth. The Soviet government has undeviatingly striven to see that space will become an arena of exclusively peaceful cooperation.
[From *Pravda,* August 1982.]

Colonel A.T. Timofeyev, Soviet Politician (1982)

48.13 All of a sudden the rock I was lying on started pitching back and forth. I could hear rock slides and boulders falling. I tried getting up, but the earth was rocking too much. I couldn't do anything; it was pitching back and forth so much.
[One of six protesters from Rocky Mountain Peace Center attempting to hike to ground zero on Pahute Mesa to stop a nuclear weapon test. Action described above took place at 1:27 P.M. Wednesday, June 25, about 3½ miles west of ground zero, during the explosion.]

Marcia Klotz, American Nuclear Protester (1986)

48.14 In the view of arms-control experts, Soviet-American reductions in nuclear arms may outweigh worry over smaller countries' arming themselves. But the experts say even an accord between the Soviet Union and the United States would leave a nuclear threat hanging over the world.

John H. Cushman, Jr., American Journalist (1986)

48.15 Anyone who has studied the ineffectual efforts of Leo Szilard, James Franck, and Albert Einstein to prevent the atomic bombing of Hiroshima can have little hope that a decisive weapon that works will not be deployed.

Jerome Grosslman, American Political Activist (1986)

48.16 Little has emerged about the effort that the USSR put into its nuclear weapons program. Yet it was an astonishing achievement. It took the Americans less than four years, starting from scratch, to make the first bomb. But American industry was advanced and unscarred by war. The corresponding Soviet effort took six years (1943–49), but Soviet industry was backward and the country was largely devastated by war. The building of a major atomic program under such conditions is testimony to Stalin's determination; to the organizational drive of Zavenyagin; and to the colossal sacrifices which must have been imposed upon the unknowing Soviet populace, whose real needs at that time must rather have been for houses and food.

Peter Kelly, English Atomic Energy Legislator (1986)

49.
Observation

Sherlock Holmes pointed out that one might see, yet not observe. That's a basic cause of much human failure.

ISAAC ASIMOV

49.1 Where there is no vision the people perish.
The Bible (circa 725 B.C.)

49.2 Science is nothing but perception.
Plato, Greek Philosopher (427 B.C.?–347 B.C.?)

49.3 Without experience nothing can be sufficiently known. For there are two modes of acquiring knowledge, namely, by reasoning and by experience. Reasoning draws a conclusion and makes us grant the conclusion, but does not make the conclusion certain . . . unless the mind discovers it by the method of experience.
Roger Bacon, English Philosopher/Clergyman (1220–1292)

49.4 Men are more apt to be mistaken in their generalizations than in their particular observations.
Niccolò Machiavelli, Italian Statesman/Political Philosopher (1469–1527)

49.5 Man, being the servant and interpreter of Nature, can do and understand so much and so much only as he has observed in fact or thought of the course of nature; beyond this he neither knows anything nor can do anything.
Francis Bacon, English Philosopher/Essayist/Statesman (1561–1626)

49.6 People who look for the first time through a microscope say now I see this and then I see that—and even a skilled observer can be fooled. On these observations I have spent more time than many will believe, but I have done them with joy, and I have taken no notice of those who have said why take so much trouble and what good is it?

Antonie van Leeuwenhoek, Dutch Scientist/Inventor (1632–1723)

49.7 While bright-eyed Science watches round.

Thomas Gray, English Poet (1716–1771)

49.8 To behold is not necessary to observe, and the power of comparing and combining is only to be obtained by education. It is much to be regretted that habits of exact observation are not cultivated in our schools; to this deficiency may be traced much of the fallacious reasoning, the false philosophy which prevails.

Alexander von Humboldt, German Naturalist/Statesman (1769–1859)

49.9 Fancy may take its flight far beyond the ken of eye or of telescope. It may expiate in the outer regions of all that is visible—and shall we have the boldness to say, that there is nothing there?

Thomas Chalmers, Scottish Theologian/Author (1780–1847)

49.10 Could a spectator exist unsustained by the earth, or any solid support, he would see around him at one view the whole contents of space—the visible constituents of the universe: and, in the absence of any means of judging their distances from him, would refer them, in the directions in which they were seen from his station, to the concave surface of an imaginary sphere, having his eye for a center, and its surface at some vast indeterminate distance.

Sir John Herschel, English Astronomer (1792–1871)

49.11 The sight of the deep-blue sky, and the clustering stars above, seems to impart a quiet to the mind.

William Cullen Bryant, American Poet/Journalist/Critic (1794–1878)

49.12 Man cannot afford to be a naturalist, to look at Nature directly, but only with the side of his eye. He must look through and beyond her, to look at her is fatal as to look at the head of Medusa. It turns the man of science to stone. I feel that I am dissipated by so many observations. I should be the magnet in the midst of all this dust and filings.

Henry David Thoreau, American Writer/Naturalist (1817–1862)

49.13 See first of all, and argue afterwards.

Jean Henri Fabre, French Entomologist/Naturalist (1823–1915)

49.14 I scrutinize life.

Jean Henri Fabre, French Entomologist/Naturalist (1823–1915)

49.15 An independent reality in the ordinary physical sense can be ascribed neither to the phenomena nor to the agencies of observation.

Niels Henrik David Bohr, Danish Physicist (1885–1962)

49.16 You can observe a lot just by watching.

Yogi Berra, American Baseball Coach/Player (1925–)

49.17 One wonders whether the rare ability to be completely attentive to, and to profit by, Nature's slightest deviation from the conduct expected of her is not the secret of the best research minds and one that explains why some men turn to most remarkably good advantage seemingly trivial accidents. Behind such attention lies an unremitting sensitivity.

Alan Gregg, English Scientist/Author (1941)

49.18 In a sense, the galaxy hardest for us to see is our own. For one thing, we are imprisoned within it, while the other galaxies can be viewed as a whole from outside. . . . Furthermore, we are far out from the center, and to make matters worse, we lie in a spiral arm clogged with dust. In other words, we are on a low roof on the outskirts of the city on a foggy day.

Isaac Asimov, American Biochemist/Author (1965)

49.19 In observation of the natural world lie the seeds of scientific innovation. . . . The inventor of the ubiquitous Velcro made his fortune by imitating in plastic the tiny hooks of a burdock.

Christopher Hallowell, American Writer (1986)

49.20 The vital act is the act of participation. "Participator" is the incontrovertible new concept given by quantum mechanics. It strikes down the term "observer" of classical theory, the man who stands safely behind the thick glass wall and watches what goes on without taking part. It can't be done, quantum mechanics says.

John Wheeler, American Physicist (1986)

49.21 At any time we can see only the objects from which light has had time to get to us since the beginning of the universe. Could I have been around when the cosmos was one second old, we could have seen only objects less than one light-second away, not as far as the moon now is. As time goes on, each observer sees objects farther and farther away.

Dietrick E. Thomsen, American Science Journalist (1987)

50. Oceanography

It is a sign of our power, and our criminal folly, that we can pollute the vast ocean and are doing so.

ISAAC ASIMOV

50.1 Multitudinous laughter of the waves of ocean.
Aeschylus, Greek Dramatist (525 B.C.?–456 B.C.?)

50.2 Nature's great and wonderful power is more demonstrated in the sea than on the land.
Pliny the Elder (Gaius Plinius Secundus), Roman Naturalist/Historian (23–79)

50.3 God quickened in the sea, and in the rivers
So many fishes of so many features
That in the waters we may see all creatures.
Even all that on the earth are to be found.
As if the world were in deep waters drowned.
For seas—as well as skies—have sun, moon, stars,
As well as air—swallows, rooks, and stares;
As well as earth—wines, roses, nettles, melons,
Mushrooms, pinks, gilliflowers, and many millions
Of other plants more rare, more strange than these,
As very fishes living in the seas;
As also rams, calves, horses, hares, and hogs,
Wolves, urchins, lions, elephants, and dogs.
Guillaume de Salluste du Bartas, French Poet/Diplomat (1544–1590)

50.4 He who ascribes the movement of the seas to the movement of the earth assumes a purely forced movement; but he who lets the seas follow the moon makes this movement in a certain way a natural one.

Johannes Kepler, German Astronomer/Mathematician (1571–1630)

50.5 The springtime sea:
all day long up-and-down,
up-and-down gently.

Taniguchi Buson, Japanese Poet/Painter (1715–1783)

50.6 The ever-raging ocean was all that presented itself to the view of this family; it irresistibly attracted my whole attention: my eyes were involuntarily directed to the horizontal line of that watery surface, which is ever in motion and ever threatening destruction to these shores . . . and who is the landman that can behold without affright so singular an element, which by its impetuosity seems to be the destroyer of this poor planet, yet at particular times accumulates the scattered fragments and produces islands and continents fit for men to dwell on!

Hector St. John de Crevecoeur, American Farmer/French Consul to New York (1735–1813)

50.7 The ocean, like the air, is the common birthright of mankind.

Thomas Jefferson, American President/Author (1743–1826)

50.8 The sea is like a silvery lake,
And o'er its calm the vessel glides
Gently as if it fear'd to wake
The slumbers of the silent tides.

Clement Clarke Moore, American Poet/Educator (1779–1863)

50.9 There is a river in the ocean.
[The Gulf Stream.]

Matthew Fontaine Maury, American Naval Officer/Oceanographer (1806–1873)

50.10 The sea from its extreme luminousness presented a wonderful and most beautiful appearance. Every part of the water which by day is seen as foam, glowed with a pale light. The vessel drove before her bows two billows of liquid phosphorus, and in her wake was a milky train. As far as the eye reached the crest of every wave was bright; and from the reflected light, the sky just above the horizon was not so utterly dark as the rest of the Heavens. It was impossible to behold this plane of matter, as if it were melted and consumed by heat, without being reminded of Milton's description of the regions of Chaos and Anarchy.

Charles Robert Darwin, English Naturalist/Evolutionist (1809–1882)

50.11 A life on the ocean wave,
A home on the rolling deep,

For the spark that nature gave
I have the right to keep.

Ambrose Bierce, American Satirist (1842–1914?)

50.12 They had neither compass, nor astronomical instruments, nor any of the appliances of our time for finding their position at sea; they could only sail by the sun, moon, and stars, and it seems incomprehensible how for days and weeks, when these were invisible, they were able to find their course through fog and bad weather; but they found it, and in the open craft of the Norwegian Vikings, with their square sails, fared north and west over the whole ocean, from Novaya Zemlya and Spitsbergen to Greenland, Baffin Bay, Newfoundland, and North America.

Fridtjof Nansen, Norwegian Explorer/Zoologist (1861–1930)

50.13 Students of oceanography from most maritime nations lament the fact that we make so little use of the countless tons of marine vegetation occurring along our coasts. Blessed as we are with a vast wealth of terrestrial resources, we have given far too little heed to the natural resources of the sea.

Claude E. Zobell, American Marine Biologist/Bacteriologist (1944)

50.14 There were tides in the earth, long before there was an ocean.

Rachel Carson, American Marine Biologist/Author (1961)

50.15 When I think of the floor of the deep sea, the single, overwhelming fact that possesses my imagination is the accumulation of sediments.

Rachel Carson, American Marine Biologist/Author (1961)

50.16 An underwater-listening device, the "hydrophone," has, in recent years, shown that sea creatures click, grunt, snap, moan, and, in general, make the ocean depths as maddeningly noisy as ever the land is.

Isaac Asimov, American Biochemist/Author (1965)

51. Optics

The rainbow, "the bridge of the gods,"

proved to be the bridge to our understanding

of light—much more important.

ISAAC ASIMOV

51.1 The light of the body is the eye.
The Bible (circa 325 A.D.)

51.2 Concerning vision alone is a separate science formed among philosophers, namely, optics, and not concerning any other sense . . . . It is possible that some other science may be more useful, but no other science has so much sweetness and beauty of utility. Therefore it is the flower of the whole philosophy and through it, and not without it, can the other sciences be known.
Roger Bacon, English Philosopher/Clergyman (1220–1292)

51.3 The first marvel of painting is that it appears detached from the wall or other flat surface, deceiving people of subtle judgment with this object that it is not separated from the wall's surface.
Leonardo da Vinci, Italian Architect/Artist/Inventor (1452–1519)

51.4 Perspective is a most subtle discovery in mathematical studies, for by means of lines it causes to appear distant that which is near, and large that which is small.
Leonardo da Vinci, Italian Architect/Artist/Inventor (1452–1519)

51.5 About ten months ago [1609] a report reached my ears that a Dutchman [Hans Lippershey] had constructed a telescope, by the aid of which visible objects, although at a great distance from the eye of the observer, were seen distinctly as if near; and some proofs of its most wonderful performances were reported, which some gave credence to, but others contradicted. A few days after, I received confirmation of the report in a letter written from Paris by a noble Frenchman, Jacques Badovere, which finally determined me to give myself to inquire into the principle of the telescope, and then to consider the means by which I might compass the invention of a similar instrument, which after a little while I succeeded in doing, through deep study of the theory of refraction; and I prepared a tube, at first of lead, in the ends of which I fitted two glass lenses, both plane on one side, but on the other side one spherically convex and the other concave.

Galileo Galilei, Italian Astronomer/Physicist (1564–1642)

51.6 Why grass is green,
or why our blood is red
Are mysteries which
none have reach'd unto.

John Donne, English Poet/Essayist/Clergyman (1572–1631)

51.7 Pictures, propagated by motion along the fibers of the optic nerves in the brain, are the cause of vision.

Isaac Newton, English Physicist/Mathematician (1642–1727)

51.8 Why has not man a microscopic eye?
For this plain reason, man is not a fly.

Alexander Pope, English Poet/Satirist (1688–1744)

51.9 I . . . formerly had two pair of spectacles, which I shifted occasionally, as in travelling I sometimes read, and often wanted to regard the prospects. Finding this change troublesome, and not always sufficiently ready, I had the glasses cut, and half of each kind associated in the same circle. . . . By this means, as I wear my spectacles constantly, I have only to move my eyes up or down, as I want to see distinctly far or near, the proper glasses being always ready.

Benjamin Franklin, American Inventor/Statesman (1706–1790)

51.10 As for what I have done as a poet, I take no pride in whatever. Excellent poets have lived at the same time with me, poets more excellent lived before me, and others will come after me. But that in my country I am the only person who knows the truth in the difficult science of colors —of that, I say, I am not a little proud, and here have a consciousness of superiority to many.

Johann Wolfgang von Goethe, German Poet/Dramatist/Novelist (1749–1832)

51.11 Where the telescope ends, the microscope begins. Which of the two has the grander view?

Victor Hugo, French Poet/Novelist/Dramatist (1802–1885)

51.12 Put by the Telescope!
Better without it man may see,
Stretch'd awful in the hush'd midnight,
The ghost of his eternity.
Coventry Patmore, English Poet (1823–1896)

51.13 Faith is a fine invention
For gentlemen who see;
But microscopes are prudent
In an emergency.
Emily Dickinson, American Poet (1830–1886)

51.14 When the Romans besieged the town (in 212 to 210 B.C.), he [Archimedes] is said to have burned their ships by concentrating on them, by means of mirrors, the sun's rays. The story is highly improbable, but is good evidence of the reputation which he had gained among his contemporaries for his knowledge of optics. At the end of this siege he was killed.
Alfred North Whitehead, English Mathematician/Philosopher (1861–1947)

51.15 If a fly had an eye like ours, the pupil would be so small that diffraction would render a clear image impossible. The only alternative is to unite a number of small and optically isolated simple eyes into a compound eye, and in the insect Nature adopts this alternative possibility.
C.J. van der Horst, Dutch Zoologist (1933)

52.
The Origin of Life

Some might accept evolution, if it allowed human beings to be created by God, but evolution won't work halfway.

ISAAC ASIMOV

52.1 God said, "Let the earth produce vegetation. . . . Let the earth produce every kind of living creature. . . ." God said, "Let us make man in our image, in the likeness of ourselves, and let them be masters of the fish of the sea, the birds of heaven, the cattle, all the wild beasts, and all the reptiles that crawl upon the earth."

The Bible **(circa 725 B.C.)**

52.2 But something else was needed, a finer being,
More capable of mind, a sage, a ruler,
So man was born, it may be, in God's image,
Or Earth, perhaps, so newly separated
From the old fire of Heaven, still retained
Some seed of the celestial force which fashioned
Gods out of living clay and running water.

Ovid (Publius Nasso Ovidius), Roman Poet (43 B.C.–17 A.D.?)

52.3 You have been considering whether the chicken came first from the egg or the egg from the chicken.

Ambrosius Theodosius Macrobius, Roman Philosopher (395–423 A.D.)

52.4 He created the heavens and the earth in accordance with the requirements of wisdom. He makes the night to cover the day, and He makes the day to cover the night; and He has pressed the sun and the moon into service; each pursues its course until an appointed time. . . . He created you from a single being; then from that He made its mate; and He has sent down for you eight head of cattle in pairs. He creates you in the wombs of your mothers, creation after creation, in threefold darkness.

The Koran (circa 650 A.D.)

52.5 Although it be a known thing subscribed by all, that the foetus assumes its origin and birth from the male and female, and consequently that the egge is produced by the cock and henne, and the chicken out of the egge, yet neither the schools of physicians nor Aristotle's discerning brain have disclosed the manner how the cock and its seed doth mint and coin the chicken out of the egge.

William Harvey, English Anatomist/Physician (1578–1657)

52.6 All things from an egg.

William Harvey, English Anatomist/Physician (1578–1657)

52.7 As for me, sir, I have not studied like you, God be thanked, and no one could boast that he had ever taught me anything; but with my small grain of common sense, my dwarfish intellect, I see things better than books can teach me, and I understand very well that this world, which we see, is not a mushroom which has sprung up in one night. I should like to ask you who has made these trees there, these rocks, this earth, and this heaven there on high, and whether all that has been made by itself?

Molière (Jean-Baptiste Poquelin), French Dramatist (1622–1673)

52.8 But the dreams about the modes of creation, enquiries whether our globe has been formed by the agency of fire or water, how many millions of years it has cost Vulcan or Neptune to produce what the fiat of the Creator would effect by a single act of will, is too idle to be worth a single hour of any man's life.

Thomas Jefferson, American President/Author (1743–1826)

52.9 Every cell from a cell.

Rudolf Virchow, German Pathologist/Statesman (1821–1902)

52.10 Every living thing from a living thing.

Louis Pasteur, French Chemist/Microbiologist (1822–1895)

52.11 Spontaneous generation is a chimera.

Louis Pasteur, French Chemist/Microbiologist (1822–1895)

52.12 Some call it evolution,
And others call it God.

William Herbert Carruth, English Poet (1859–1924)

52.13 When ultra-violet light acts on a mixture of water, carbon dioxide, and ammonia, a vast variety of organic substances are made, including sugars and apparently some of the materials from which proteins are built up. . . . before the origin of life they must have accumulated till the primitive oceans reached the consistency of hot dilute soup. . . . The first living or half-living things were probably large molecules synthesized under the influence of the sun's radiation, and only capable of reproduction in the particularly favorable medium in which they originated. . . .

It is probable that all organisms now alive are descended from one ancestor, for the following reason. Most of our structural molecules are asymmetrical, as shown by the fact that they rotate the plane of polarized light, and often form asymmetrical crystals. But of the two possible types of any such molecule, related to one another like a right and left boot, only one is found throughout living nature. The apparent exceptions to this rule are all small molecules which are not used in the building of the large structures which display the phenomena of life.

John Burdon Sanderson Haldane, English Geneticist (1892–1964)

52.14 The first organic substances which had hitherto remained in the atmosphere were now dissolved in the water and fell to the ground with it. . . . When these substances fell from the atmosphere into the primeval ocean they did not stop interacting with one another. Individual components of organic substances floating in the water met and combined with one another. . . .

Substances with large and complicated particles have a great tendency to form colloidal solutions in water. . . . For various, sometimes extremely slight causes, the dissolved substances come out of the colloidal solution in the form of precipitates, coagula, or gels. It is impossible, incredible, to suppose that in the course of many hundreds or even thousands of years during which the terrestrial globe existed, the conditions did not arise "by chance" somewhere which would lead to the formation of a gel in a colloidal solution. . . . With certain reservations we can even consider that first piece of organic slime which came into being on the Earth as being the first organism.

A.I. Oparin, Russian Scientist (1924)

52.15 Almost in the beginning was curiosity.

Isaac Asimov, American Biochemist/Author (1965)

52.16 Considering the difficulties represented by the lack of water, by extremes of temperature, by the full force of gravity unmitigated by the buoyancy of water, it must be understood that the spread to land of life forms that evolved to meet the conditions of the ocean represented the greatest single victory won by life over the inanimate environment.

Isaac Asimov, American Biochemist/Author (1965)

52.17 It seems that the absence of free water for any prolonged period may prove to be the decisive obstacle in the path of evolution. At least one of the greatest tragedies of our Universe is thirst, which may prove to prevent the blossoming-out of life on innumerable celestial bodies, which are considerably smaller than our planet, but seem to have had otherwise suitable conditions for the propagation of organisms.

G. Mueller, Writer (1967)

52.18 The question of the origin of life is essentially speculative. We have to construct, by straightforward thinking on the basis of very few factual observations, a plausible and self-consistent picture of a process which must have occurred before any of the forms which are known to us in the fossil record could have existed.

J.D. Bernal, English Crystallographer (1967)

52.19 We suppose that on the primitive Earth there was land including rocks that weathered to create supersaturated solutions from which clays crystallised from time to time. More specifically, we might imagine such solutions percolating through porous rock or a bed of sand with clay crystallisation being nucleated more or less at random—say on the surfaces of the surrounding rocks—giving rise to arbitrary substitution patterns. In at least some of these clays, however, these initial arbitrary patterns are subsequently printed off again and again through crystal growth. We see this as the pre-condition for life: genetic crystals in a supersaturated solution automatically printing off any pattern that they just happen to hold.

A. Graham Cairns-Smith, Canadian Chemistry Professor (1971)

52.20 All cells generate energy from sunlight or from energy-rich food molecules. They all produce the energy-rich molecule adenosine triphosphate (ATP), used to power energy-requiring reactions. All contain replicating systems based on the molecules DNA and RNA, without which reproduction and growth would be impossible. The universality of these molecules in all organisms living today tells us that all life on Earth is ultimately related.

Lynn Margulis, American Biologist/Author (1982)

52.21 Some two billion years ago, cyanobacteria (probably the first organisms to generate oxygen photosynthetically) made drastic changes in the atmosphere. It is doubtful that any organisms since then have had such a profound effect on the planet.

Lynn Margulis, American Biologist/Author (1982)

52.22 Recently, we've reported that we have made all five bases, the compounds that spell out the instructions for all life and are a part of the nucleic acids, RNA and DNA. Not only did we make all five bases but we found them in a meteorite! So that these two things coming together really assure us that the molecules necessary for life can be found in the absence of life. This was the biggest stumbling block.

Cyril Ponnamperuma, American Chemist (1985)

52.23 If we assume life begins from self-assembly, then the nonbiological self-assembly seen with chemicals from inside the Murchison meteorite shows how the essential membrane of the first microorganism might have formed.
David W. Deamer, American Biophysicist (1986)

52.24 The general feeling still is that it all began in water.
André Brack, French Chemist (1986)

53. Ornithology

Birds sing sweetly, but someone awakened by them at 5 A.M. of a summer morning might dispute the adverb.

ISAAC ASIMOV

53.1 He is a fool who lets slip a bird in the hand for a bird in the bush.
Plutarch, Greek Biographer/Essayist (46–120)

53.2 A rare bird on earth, like a black swan.
Juvenal (Decimus Junius Juvenalis), Roman Poet/Satirist (60?–127)

53.3 In one of the churches of Clermont-Ferrand, while early-morning matins were being celebrated on some feast-day or other, a bird called a crested lark flew in, spread its wings over all the lamps which were shining and put them out so quickly that you would have thought that someone had seized hold of them all at once and dropped them into a pool of water. It then flew into the sacristy, under the curtain, and tried to extinguish the candle there, but the vergers managed to catch it and they killed it. In the same way another bird put out the lamps lighted in Saint Andrew's church.
Gregory of Tours, Frankish Ecclesiastic/Historian (538–594)

53.4 Twice or thrice the young bird may be deceived, but before the eyes of the full-fledged it is vain to spread the net, or speed the arrow.
Dante Alighieri, Italian Poet (1265–1321)

53.5 How lavish nature has adorn'd the year
How the pale primrose and blue violet spring,
And birds essay their throats disus'd to sing.
Geoffrey Chaucer, English Poet (1340?–1400)

53.6 Birds of a feather will flock together.
John Minshew, English Lexicographer (1550?–1627?)

53.7 When daisies pied and violets blue
And lady-smocks all silver-white
And cuckoo-buds of yellow hue
Do paint the meadows with delight,
The cuckoo then, on every tree,
Mocks married men; for thus sings he, "Cuckoo!
Cuckoo, cuckoo!" O word of fear,
Unpleasing to a married ear!
William Shakespeare, English Dramatist/Poet (1564–1616)

53.8 See how that pair of billing doves
With open murmurs own their loves
And, heedless of censorious eyes,
Pursue their unpolluted joys:
No fears of future want molest
The downy quiet of their nest.
Lady Mary Wortley Montagu, English Poet/Author (1689–1762)

53.9 I would pick from you,
if it were a blossom spray—
one trill, cuckoo!
Kodo, Japanese Poet (?–1738)

53.10 A good ornithologist should be able to distinguish birds by their air as well as by their colors and shape; on the ground as well as on the wing, and in the bush as well as in the hand. For, though it must not be said that every species of birds has a manner peculiar to itself, yet there is somewhat, in most genera at least, that at first sight discriminates them and enables a judicious observer to pronounce upon them with some certainty.
Gilbert White, English Naturalist/Ecologist (1720–1793)

53.11 When it [the hummingbird] feeds, it appears as if immovable, though continually on the wing; and sometimes, from what motives I know not, it will tear and lacerate flowers into a hundred pieces, for, strange to tell, they are the most irascible of the feathered tribe. Where do passions find room in so diminutive a body? They often fight with the fury of lions until one of the combatants falls a sacrifice and dies.
Hector St. John de Crevecoeur, American Farmer/French Consul to New York (1735–1813)

53.12 I have heard the nightingale in all its perfection, and I do not hesitate to pronounce that in America it would be deemed a bird of the third rank only, our mocking-bird and fox-colored thrush being unquestionably superior to it.
Thomas Jefferson, American President/Author (1743–1826)

53.13 Art thou the bird whom Man loves best,
The pious bird with the scarlet breast,
Our little English Robin;
The bird that comes about our doors
When autumn winds are sobbing?
William Wordsworth, English Poet (1770–1850)

53.14 The swallow is come!
The swallow is come!
O, fair are the seasons, and light
Are the days that she brings,
With her dusky wings,
And her bosom snowy white!
Henry Wadsworth Longfellow, American Poet (1807–1882)

53.15 The crack-brained bobolink courts his crazy mate,
Posed on a bulrush tipsy with his weight.
Oliver Wendell Holmes, American Physician/Poet/Author (1809–1894)

53.16 Just after sundown I see a large flock of wild geese in a perfect harrow cleaving their way toward the northeast, with Napoleonic tactics splitting the forces of winter.
Henry David Thoreau, American Writer/Naturalist (1817–1862)

53.17 At a distance in the meadow I hear still, at long intervals, the hurried commencement of the bobolink's strain, the bird just dashing into song, which is as suddenly checked, as it were, by the warder of the seasons, and the strain is left incomplete forever. Like human beings they are inspired to sing only for a short season.
Henry David Thoreau, American Writer/Naturalist (1817–1862)

53.18 I hear the scream of a great hawk, sailing with a ragged wing against the high wood-side, apparently to scare his prey and so detect it—shrill, harsh, fitted to excite terror in sparrows and to issue from his split and curved bill. I see his open bill the while against the sky. Spit with force from his mouth with an undulatory quaver imparted to it from his wings or motion as he flies.
Henry David Thoreau, American Writer/Naturalist (1817–1862)

53.19 Noise proves nothing. Often a hen who has merely laid an egg cackles as if she laid an asteroid.
Mark Twain (Samuel Langhorne Clemens), American Journalist/ Novelist (1835–1910)

53.20 A hen is only an egg's way of making another egg.
Samuel Butler, English Novelist (1835–1902)

53.21 Our nights are days to the owls. While we sleep under domestic quilts, they are wide awake flying by hedgerows or over waste places where the poppies appear black in the moonlight, cognisant of the slightest stir, of the faintest shadow, even stooping for a beetle.

Llewelyn Powys, English Author (1884–1939)

53.22 In order to see birds it is necessary to become a part of the silence. One has to sit still like a mystic and wait. One soon learns that fussing, instead of achieving things, merely prevents things from happening.

Robert Staughton Lynd, American Sociologist (1892–1970)

53.23 The Bower birds of Australia and New Guinea build bowers for courtship with brightly colored fruits or flowers which are not eaten but left for display and replaced when they wither. . . . They stick to a particular color scheme. Thus a bird using blue flowers will throw away a yellow flower inserted by the experimenter, while a bird using yellow flowers will not tolerate a blue one.

W.H. Thorpe, English Animal Behaviorist/Educator (1902–)

53.24 Can anyone imagine a more humiliating experience than having "the village cop" tell ornithologists "what kind of bird that is"?

Julian K. Potter, English Ornithologist (1943)

53.25 Navigators should observe the bird life, for deductions may often be drawn from the presence of certain species. Shags are a sure sign of the close proximity of land. . . . The snow petrel is invariably associated with ice and is of great interest to mariners as an augury of ice conditions in their course. . . .

***The United States Pilot,* Government Navigational Guide to Antarctica (1960)**

53.26 None of these specializations is inherently incompatible with egg retention or viviparity. Rather, these specializations have greatly diminished the potential advantages of egg retention while magnifying the associated disadvantages, such as loss of fecundity, increased maternal mortality, and decreased paternal investment.

Daniel Blackburn and Howard Evans, American Science Researchers (1986)

53.27 Delaware Bay is unquestionably the largest spring staging site for shorebirds in eastern North America. No other spot comes close. Together with the Copper River Delta in Alaska, Gray's harbor in Washington, the Bay of Fundy, and Cheyenne Bottoms, Kansas, each of which harbors hundreds of thousands to millions of shorebirds during northbound or southbound migrations, Delaware Bay underpins the entire migration system of New World shorebirds. Without these sites, the stupendous migrations undertaken by North America's shorebirds would be impossible.

J.P. Myers, American Zoologist/Ornithologist (1986)

53.28 Enormous quantities of nuts are moved by jays each fall. . . . Caches never retrieved by birds and undiscovered by mammals lead to the regeneration of trees and the eventual perpetuation of nuts for future generations of blue jays. . . .

Thus, the distribution of oaks in nature may mirror the collective behavioral decisions made by the community of jays as they select, disperse, cache, and retrieve nuts.

W. Carter Johnson and Curtis S. Adkinson, American Botanists/ Biologists (1986)

54. Paleontology

Of all extinct life-forms, dinosaurs are the most popular. Why that should be is not clear.

ISAAC ASIMOV

54.1 It seems wonderful to everyone that sometimes stones are found that have figures of animals inside and outside. For outside they have an outline, and when they are broken open, the shapes of the internal organs are found inside. And Avicenna says that the cause of this is that animals, just as they are, are sometimes changed into stones.

Albertus Magnus, German Clergyman/Theologian/Scholar (1200?–1280)

54.2 Whence we see spiders, flies, or ants, entombed and preserved for ever in amber, a more than royal tomb.

Francis Bacon, English Philosopher/Essayist/Statesman (1561–1626)

54.3 As soon as the sea breeze came, I took a walk towards the shores of a river [the Potomac]. As I was searching for the most convenient spot to descend to the shores, I perceived a large flat stone lying on the ground. As they are very scarce in this part of the country, I stopped to view it and to consider whether it had not been left there on some peculiar account. On looking at it more attentively, I perceived the marks of ancient sea-shells encrusted on its surface. How could this stone have received these marine impressions? How could it be brought here where stones are so scarce?

Hector St. John de Crevecoeur, American Farmer/French Consul to New York (1735–1813)

54.4 It is well known, that on the Ohio, and in many parts of America further north, tusks, grinders, and skeletons of unparalleled magnitude are found in great numbers, some lying on the surface of the earth, and some a little below it. . . . But to whatever animal we ascribe these remains, it is certain that such a one has existed in America, and that it has been the largest of all terrestrial beings.

Thomas Jefferson, American President/Author (1743–1826)

54.5 Those enormous bones which are found in great quantity in Ohio—the exact knowledge of those objects is more important toward the theory of the earth, than is generally thought.

Jean-Baptiste de Monet, Chevalier de Lamarck, French Naturalist (1744–1829)

54.6 A controversy was maintained for more than a century respecting the origin of fossil shells and bones—were they organic or inorganic substances? That the latter opinion should for a long time have prevailed, and that these bodies should have been supposed to be fashioned into their present form by a plastic virtue, or some other mysterious agency, may appear absurd; but it was, perhaps as reasonable a conjecture as could be expected from those who did not appeal, in the first instance, to the analogy of living creation, as affording the only source of authentic information.

Sir Charles Lyell, Scottish Geologist (1797–1875)

54.7 Are God and Nature then at strife,
That Nature lends such evil dreams?
So careful of the type she seems,
So careless of the single life . . .
"So careful of the type," but no.
From scarped cliff and quarried stone
She cries, "A thousand types are gone;
I care for nothing, all shall go."

Alfred, Lord Tennyson, English Poet (1809–1892)

54.8 Was the oldest *Homo sapiens* Pliocene or Miocene? In still older strata do the fossilized bones of an ape, more anthropoid, or a man, more pithecoid than any yet known await the researches of some unborn paleontologist?

Thomas Henry Huxley, English Biologist/Evolutionist (1825–1895)

54.9 We have only indirect means of knowing the courage and activity of the Neanderthals in the chase, through the bones of animals hunted for food which are found intermingled with the flints around their ancient hearths.

Henry Fairfield Osborn, American Paleontologist/Zoologist (1857–1935)

54.10 The Chinese . . . use fossil teeth as one of their principal medicines. Some Chinese families have for centuries been in the business of "mining" fossils to supply the drug trade.

Robert Broom, Scotch/South African Paleontologist (1866–1951)

54.11 To a certain extent function is reflected in the [fossil] bones. But you cannot solve this puzzle by looking only at the bones, for the answer is not there. The answer lies in how the man or the animal functioned, and that lies in the experimental laboratory where you can study comparable parts.

Sherwood L. Washburn, American Paleontologist/Physiologist (1911–1980)

54.12 Behold the mighty dinosaur
Famous in prehistoric lore,
Not only for his weight and length
But for his intellectual strength.
You will observe by these remains
The creature had two sets of brains—
One in his head (the usual place),
The other at his spinal base.
Thus he could reason *a priori*
As well as *a posteriori*.
No problem bothered him a bit
He made both head and tail of it.

So wise was he, so wise and solemn,
Each thought filled just a spinal column.
If one brain found the pressure strong
It passed a few ideas along.
If something slipped his forward mind
'Twas rescued by the one behind.
And if in error he was caught
He had a saving afterthought.
As he thought twice before he spoke
He had no judgment to revoke.
Thus he could think without congestion
Upon both sides of every question.
Oh, gaze upon this model beast
Defunct ten million years at least.

Bert Liston Taylor, American Journalist/Humorist (1920)

54.13 Why the dinosaurs died out is not known, but it is supposed to be because they had minute brains and devoted themselves to the growth of weapons of offense in the shape of numerous horns. However that may be, it was not through their line that life developed.

Bertrand Russell, English Philosopher/Mathematician (1952)

54.14 The creation model, on the other hand, must interpret the [geological] column in terms of essentially continuous deposition, all accomplished in a relative short time—not instantaneously, of course, but over a period of months or years, rather than millions of years. In effect, this means that the organisms represented in the fossil record must all have been living contemporaneously, rather than scattered in separate time-frames over hundreds of millions of years.

H.M. Morris, American Scientific Creationist/Author (1974)

54.15 Conventional science—uniformitarianism—does not allow catastrophesto occur on Earth on a grand scale. Yet the evidence presented by the dinosaur and mammoth remains can most readily be explained by just such catastrophes. For the mammoths, the evidence for a large-scale catastrophe, or a series of catastrophes, is so great that any other explanation is woefully inadequate. For the dinosaurs, other explanations seem plausible—until one is faced with the fact that it was not just the dinosaurs that disappeared, but that between one-half and three-quarters of all living species throughout the world became extinct at the same time.

Peter Warlow, British Physicist/Author/Lecturer (1982)

54.16 In a diatomaceous earth quarry at Lompoc, California, the fossil skeleton of a whale was found in 1976. The whale is standing on end, with its head uppermost.

Peter Warlow, British Physicist/Author/Lecturer (1982)

54.17 The rate of extinction is now about 400 times that recorded through recent geological time and is accelerating rapidly. Under the best of conditions, the reduction of diversity seems destined to approach that of the great natural catastrophes at the end of the Paleozoic and Mesozoic Eras, in other words, the most extreme for 65 million years. And in at least one respect, this human-made hecatomb is worse than any time in the geological past. In the earlier mass extinctions . . . most of the plant diversity survived; now, for the first time, it is being mostly destroyed.

Edward O. Wilson, American Entomologist/Sociobiologist (1985)

54.18 [Dinosaurs are] a totally unique form of life that this planet has never seen since.

Stephen Czerkas, American Paleontologist/Artist (1986)

54.19 We have very strong physical and chemical evidence for a large impact; this is the most firmly established part of the whole story. There is an unquestionable mass extinction at this time, and in the fossil groups for which we have the best record, the extinction coincides with the impact to a precision of a centimeter or better in the stratigraphic record. This exact coincidence in timing strongly argues for a causal relationship.
[Theory held by himself and his father, physicist Luis W. Alvarez, that dinosaurs died out suddenly after the earth was hit by a 6-mile-wide asteroid or comet.]

Walter Alvarez, American Geologist (1986)

54.20 We now believe that *Gigantopithecus* became extinct during the mid-Pleistocene, perhaps surviving until about 300,000 years ago, or even a bit later. Some of our Chinese colleagues would say, in fact, that some form of *Gigantopithecus* might have survived to this day as the so-called yeti, or abominable snowman.

Dr. John W. Olsen, American Anthropologist (1986)

54.21 *Archaeopteryx* is not a forgery.
[*Archaeopteryx lithographica* is the oldest species of fossil bird and is an example of organic evolution.]

Alan J. Charig, American Naturalist (1986)

54.22 Often of dwarf size, elephants, deer, and hippos, along with other unique local species, once inhabited many Mediterranean islands. . . . The animals that lived on widely separated islands were often similar, reflecting parallel events in the founding of island populations and their subsequent evolution.

Paul Y. Sondaar, Dutch Paleontologist/Educator (1986)

54.23 The possibility that these people, like other island species, were endemic to Sardinia needs to be studied; the fossils show that they were anatomically different from contemporaneous humans from the mainland.
[On human fossil remains found in 1985, dated at about 14,000 years old.]

Paul Y. Sondaar, Dutch Paleontologist/Educator (1986)

54.24 What happened to those Ice Age beasts? What caused the mammoth and mastodon and wooly rhinoceros to pay the ultimate Darwinian penalty, while bison and musk ox survived? Why didn't the fauna of Africa suffer the kinds of losses evident in other regions of the world? And if something like climatic change caused the extinction of North America's Pleistocene horse, how have feral horses managed to reestablish themselves on the western range?

John R. Alden, American Anthropologist/Author (1986)

54.25 Throughout most of the world at that crucial time, many groups of plants and animals—from the giant reptiles, such as dinosaurs, to the intricately shelled sea creatures, the ammonites, to the microscopic, one-celled marine plants—suddenly became extinct. . . .

The wholesale extinctions that occurred in Europe and North America were not in evidence at the extreme southern latitudes. Information from Seymour Island about the extinction indicates that, whatever the cause, the result was not as severe in the southern regions as in other areas of the earth.

William J. Zinmeister, American Paleontologist/Geoscientist (1986)

54.26 Sharp declines in the number and variety of species alternate with rapid radiations, and there are occasional major extinctions. . . .

The pattern of these declines and radiations exactly matches the changing rate of sea-floor spreading in the Atlantic and Pacific Oceans. . . . A period of enhanced tectonic activity lasted from 73 to 65 million years ago, at the end of the Cretaceous, and coincided with the great extinctions, including the death of the dinosaurs.

Jeff Hecht, American Science Journalist (1986)

54.27 By looking in aquatic assemblages, we may be hunting in the wrong places for the ancestors of land organisms.

Gregory J. Retallack, American Paleontologist (1987)

55. Parapsychology

It is a great deal easier to believe in

the existence of parapsychological phenomena,

if one is ignorant of, or indifferent to,

the nature of scientific evidence.

ISAAC ASIMOV

55.1 The best of seers is he who guesses well.
Euripedes, Greek Dramatist (480 B.C.–406 B.C.)

55.2 I wonder that a soothsayer doesn't laugh whenever he sees another soothsayer.
Marcus Tullius Cicero, Roman Orator/Statesman (106 B.C.–43 B.C.)

55.3 Oct. 26, [1635] Monday, this morning between four and five of the clock, lying at Hampton Court, I dreamed that I was going out in haste, and that when I came into my outer chamber, there was my servant Will Pennell in the same riding suit which he had on that day sennight at Hampton Court with me. Methought I wondered to see him—for I left him sick at home—and asked him how he did and what he made there. And that he answered me, he came to receive my blessing; and with that fell on his knees. That hereupon I laid my hand on his head, and prayed over him, and therewith awaked. When I was up, I told this to them in my chamber; and added that I should find Pennell dead or dying. My coach came; and when I came home, I found him past sense and giving up the ghost. So my prayers, as they had frequently before, commended him to God.
William Laud, English Archbishop (1573–1645)

55.4 Pataphysics is the science of imaginary solutions, which symbolically attributes the properties of objects, described by their virtuality, to their lineaments.

Alfred Jarry, French Playwright/Novelist (1873–1907)

55.5 I resolved to obtain from myself [through automatic writing] what we were trying to obtain from them, namely a monologue spoken as rapidly as possible without any intervention on the part of the critical faculties, a monologue consequently unencumbered by the slightest inhibition and which was, as closely as possible akin to *spoken thought*. It had seemed to me, and still does . . . that the speed of thought does not necessarily defy language, nor even the fast-moving pen.

André Breton, French Writer/Poet (1896–1966)

55.6 Extrasensory perception is a scientifically inept term. By suggesting that forms of human perception exist beyond the senses, it prejudges the question.

Margaret Mead, American Anthropologist (1901–1978)

55.7 This is an acceptable significant result in the other sciences, but parapsychology is more conservative and considers such odds as only suggestive of extra chance performance.
[Commenting on the results of ESP experiments carried out during the Apollo 14 space flight in which two of the four subjects on Earth correctly identified cards he held in 51 out of 200 tries.]

Commander Edgar D. Mitchell, Jr., American Astronaut (1971)

55.8 I have learned anyway, from research on the growing band who bend metal and move pendulums in apparently paranormal ways, that it is seldom possible to establish very much more than that these things do happen, sometimes, usually when you least expect it, but seldom in a way that will carry sufficient authority to convince anyone who requires very rigid mechanical demonstrations of reality.

Lyall Watson, American Biologist (1979)

55.9 What separates parapsychology from occultism is not so much the subject matter, which may even be the same, but the *methodology.*

Rhea White, American Parapsychologist (1981)

55.10 Just how striking, unusual, or improbable must a coincidence be before one accepts the even more improbable explanation that a paranormal form of communication is involved?

Ruth Reinsel, American Writer (1982)

55.11 It is not the fact that events cannot be explained by science that will convince skeptics of the existence of hitherto unnoticed "underlying order." Something has to convince us that that failure of explanation is worthy of serious notice.

Galen K. Fletcher, American Writer (1982)

55.12 It constantly confounds me that not only the young, but also many certified intellectuals accept uncritically the superiority of spontaneous or unconscious products of mind [such as automatic writing] over those subjected to conscious, rational control.

Roger Shattuck, American Writer (1984)

56.
Physics

Physics is the basic science. One can easily argue that all other sciences are specialized aspects of physics.

ISAAC ASIMOV

56.1 Give me a firm spot on which to stand, and I will move the earth [with the lever].
Archimedes, Greek Mathematician/Inventor (287 B.C.–212 B.C.)

56.2 Physics inquires whether the world is eternal, or perpetual, or had a beginning and will have an end in time, or whether none of these alternatives is accurate.
John of Salisbury, English Churchman (1115–1180)

56.3 A vacuum is repugnant to reason.
René Descartes, French Mathematician/Philosopher (1596–1650)

56.4 As to a perfect science of natural bodies . . . we are, I think, so far from being capable of any such thing that I conclude it lost labor to seek after it.
John Locke, English Philosopher (1632–1704)

56.5 Vacuum I call every place in which a body is able to move without resistance.
Isaac Newton, English Physicist/Mathematician (1642–1727)

56.6 Every body continues in its state of rest or of uniform motion in a straight line, except in so far as it is compelled to change that state by forces impressed upon it.
Isaac Newton, English Physicist/Mathematician (1642–1727)

56.7 Now, that this whiteness is a Mixture of the severaly color'd rays, falling confusedly on the paper, I see no reason to doubt of.
[From Isaac Newton's "Answer to some Considerations (of Robert Hooke) upon his Doctrine of Light and Colors," in *Philosophical Transactions,* November 18, 1672.]

Isaac Newton, English Physicist/Mathematician (1642–1727)

56.8 And first, I suppose, that there is diffused through all places an ethereal substance, capable of contraction and dilation, strongly elastic, and, in a word, much like air in all respects, but far more subtle.
2. I suppose this ether pervades all gross bodies, but yet so as to stand rarer in their pores than in free spaces, and so much the rarer, as their pores are less.
3. I suppose the rarer ether within bodies, and the denser without them; not to be terminated in a mathematical superficies, but to grow gradually into one another.
[Letter of Newton to Robert Boyle, February 28, 1678.]

Isaac Newton, English Physicist/Mathematician (1642–1727)

56.9 We may fairly claim for the study of Physics the recognition that it answers to an impulse implanted by nature in the constitution of man.

John Tyndall, English Physicist (1820–1893)

56.10 Negatively charged particles can be torn from them [atoms] by the action of electrical forces.

Sir Joseph John Thomson, English Physicist (1856–1940)

56.11 $E=hf$
[Formula describing the energy of a photon, or quantum theory, E=Energy, h=Planck's constant, f=frequency of light.]

Max Planck, German Physicist (1858–1947)

56.12 The most incomprehensible thing about the world is that it is comprehensible.

Albert Einstein, German/American Physicist (1879–1955)

56.13 Our new idea is simple: to build a physics valid for all coordinate systems.

Albert Einstein, German/American Physicist (1879–1955)

56.14 The more success the quantum theory has, the sillier it looks.

Albert Einstein, German/American Physicist (1879–1955)

56.15 Physics is the nature of the case indeterminate, and therefore the affair of statistics.

Max Born, German/British Physicist (1882–1970)

56.16 Physics regards matter solely as a vehicle of energy . . . physics may be regarded as the science of energy, precisely as chemistry may be regarded as the science of matter.

G.F. Barker, English Physicist (1892–)

56.17 The science of physics, in addition to the general laws of dynamics and their application to the interaction of solid, liquid, and gaseous bodies, embraces the theory of those agents which were formerly designated as imponderables—light, heat, electricity, magnetism, etc.; and all these are now treated as forms of motion, manifestations of the same fundamental energy.

John B. Stallo, American Physicist (1900)

56.18 Physics is becoming so unbelievably complex that it is taking longer and longer to train a physicist. It is taking so long, in fact, to train a physicist to the place where he understands the nature of physical problems that he is already too old to solve them.

Eugene Paul Wigner, Hungarian/American Educator (1902–)

56.19 The physicists have known sin; and this is a knowledge which they cannot lose.

J. Robert Oppenheimer, American Physicist (1904–1967)

56.20 A happy era of physics that will not come again.
[Referring to the height of activity at the Bohr Institute.]

Hendrik Brugt Gerhard Casimir, Dutch Physicist (1909–)

56.21 There is no democracy in physics. We can't say that some second-rate guy has as much right to an opinion as Fermi.

Luis Walter Alvarez, American Physicist (1911–)

56.22 The work of Planck and Einstein proved that light behaved as particles in some ways and that the ether therefore was not needed for light to travel through a vacuum. When this was done, the ether was no longer useful and it was dropped with a glad cry. The ether has never been required since. It does not exist now; in fact, it never existed.

Isaac Asimov, American Biochemist/Author (1976)

56.23 If ever an equation has come into its own it is Einstein's $e = mc^2$. Everyone can rattle it off now, from the highest to the lowest.

Isaac Asimov, American Biochemist/Author (1976)

56.24 "Wu Li" was more than poetic. It was the best definition of physics that the conference would produce. It caught that certain something, that living quality that we were seeking to express in a book, that thing without which physics becomes sterile. "Wu" can mean either "matter" or "energy." "Li" is a richly poetic word. It means "universal order" or "universal law." It also means "organic patterns." The grain in a panel of wood is Li. The organic pattern on the surface of a leaf is also Li, and so is the texture of a rose petal. In short, Wu Li, the Chinese word for physics, means "patterns of organic energy" ("matter/energy" [Wu] + "universal order/organic patterns" [Li]). This is remarkable since it reflects a world view which the founders of western science (Galileo and Newton) simply did not comprehend, but toward which virtually every physical theory of import in the twentieth century is pointing!

Gary Zukav, American Author (1979)

56.25 Gravity has given modern theorists a problem. As physicists have tried to combine all the known forces of nature—the gravitational, electromagnetic, weak, and strong forces—into one unified field theory, they have been unable to incorporate gravity without postulating the existence of other, as-yet-undiscovered forces. . . . In particular, the unified field theories say that the force of attraction between two bodies is given by Newton's formula, plus a much smaller "fifth force" that seems to come into play at distances of about 100 to 1,000 meters.

Stefi Weisburd, American Science Journalist (1987)

56.26 Experiments conducted deep in a South Dakota gold mine have reportedly produced evidence for an extremely rare form of radioactive decay whose existence, if proved, would overthrow one of the basic laws of physics.

Known as the law of lepton conservation, it says the total number of lightweight or weightless particles—leptons—emerging from an atomic reaction must equal the number entering into it.

Walter Sullivan, American Science Journalist (1987)

57. Physiology

The living human being seems to consist

of nothing more than matter and energy.

Spirit is merely an assumption.

ISAAC ASIMOV

57.1 What tent poles are to tents, and walls to houses, so to animals their bony structure; the other parts adapt themselves to this and change with it.
Claudius Galen, Greek Physician/Scholar (130–200)

57.2 "True is it, my incorporate friends," quoth he,
"That I receive the general food at first,
Which you do live upon; and fit it is,
Because I am the storehouse and the shop
Of the whole body. But, if you do remember,
I send it through the rivers of your blood,
Even to the court, the heart, to th' seat o' th' brain;
And, through the cranks and offices of man,
The strongest nerves and small inferior veins
From me receive that natural competency
Whereby they live. And though that all at once . . . cannot
See what I do deliver out to each
Yet I can make my audit up, that all
From me do back receive the flour of all,
And leave me but the bran.
[The belly's answer to a complaint from the other members of the body that it received all the food but did no work.]
William Shakespeare, English Dramatist/Poet (1564–1616)

57.3 We conclude that blood lives of itself and that it depends in no ways upon any parts of the body. Blood is the cause not only of life in general, but also of longer or shorter life, of sleep and waking, of genius, aptitude, and strength. It is the first to live and the last to die.

William Harvey, English Anatomist/Physician (1578–1657)

57.4 To vary the compression of the muscle therefore, and so to swell and shrink it, there needs nothing but to change the consistency of the included ether. . . . Thus may therefore the soul, by determining this ethereal animal spirit or wind into this or that nerve, perhaps with as much ease as air is moved in open spaces, cause all the motions we see in animals.
[From Newton's Second Paper on Color and Light, read at the Royal Society in 1675.]

Isaac Newton, English Physicist/Mathematician (1642–1727)

57.5 In cold countries the aqueous particles of the blood are exhaled slightly by perspiration; they remain in great abundance. One can therefore make use of spirituous liquors without the blood coagulating. It is full of humors. Strong liquors, which give movement to the blood, may be suitable there.

Baron de la Brède et de Montesquieu, French Philosopher/Author (1689–1755)

57.6 Bile makes man passionate and sick; but without bile man could not live.

François-Marie Arouet de Voltaire, French Author/Philosopher (1694–1778)

57.7 No knowledge can be more satisfactory to a man than that of his own frame, parts, their functions and actions.

Thomas Jefferson, American President/Author (1743–1826)

57.8 This pure species of air [oxygen] has the property of combining with the blood and . . . this combination constitutes its red color.

Antoine Lavoisier, French Chemist (1743–1794)

57.9 It is not the organs—that is, the character and form of the animal's bodily parts—that have given rise to its habits and particular structures. It is the habits and manner of life and the conditions in which its ancestors lived that have in the course of time fashioned its bodily form, its organs and qualities.

Jean-Baptiste de Monet, Chevalier de Lamarck, French Naturalist (1744–1829)

57.10 The entire human body is disposed for a vertical posture.

Georges Cuvier, French Zoologist/Anatomist (1769–1832)

57.11 The organism, as we have long been saying, is a little world (microcosm) in the great universe (macrocosm).

Claude Bernard, French Physiologist (1813–1878)

57.12 Physiology is the experimental science *par excellence* of all sciences; that in which there is least to be learnt by mere observation, and that which affords the greatest field for the exercise of those faculties which characterize the experimental philosopher.

Thomas Henry Huxley, English Biologist/Evolutionist (1825–1895)

57.13 Connected by innumerable ties with abstract science, Physiology is yet in the most intimate relation with humanity; and by teaching us that law and order, and a definite scheme of development, regulate even the strangest and wildest manifestations of individual life, she prepares the student to look for a goal even amidst the erratic wanderings of mankind, and to believe that history offers something more than an entertaining chaos—a journal of a toilsome, tragi-comic march nowither.

Thomas Henry Huxley, English Biologist/Evolutionist (1825–1895)

57.14 The soul of man is—objectively considered—essentially similar to that of all other vertebrates; it is the physiological action or function of the brain.

Ernst Heinrich Haeckel, German Biologist/Philosopher (1834–1919)

57.15 All explanations of a materialistic kind the physiologist finds soon to be highly incomplete, since they touch only one single side of life; and he comes to see, above everything, that life cannot be explained from something else, but must be conceived and understood in itself.

E.S. Russell, English Scientist (1930)

57.16 There is no question but that man's heart outperforms all other hearts in existence. (The tortoise's heart may last longer but it lives nowhere near as intensely.) Why man should be so long-lived is not known, but man, being what he is, is far more interested in asking why he does not live still longer.

Isaac Asimov, American Biochemist/Author (1965)

57.17 Our brains probably have natural counterparts for just about any drug you can name.

Candace Pert, American Neuroscientist (1984)

57.18 We found that each disconnected hemisphere [of the brain] was capable of sustaining its own conscious awareness, each largely oblivious of experiences of the other.

Roger Wolcott Sperry, American Neurophysiologist/Psychobiologist (1984)

57.19 The mind can quickly scan not only the past, but also the projected future consequences of a choice. Its dynamics transcend the time and space of brain physiology.

Roger Wolcott Sperry, American Neurophysiologist/Psychobiologist (1984)

57.20 The conscious mind doesn't initiate voluntary actions. I propose that the performance of every conscious voluntary act is preceded by special unconscious cerebral processes that begin about one-half second or so before the act.

Benjamin Libet, American Physiologist (1986)

57.21 It looks as if the forebrain literally develops under the influence of the nose. . . . For instance human babies born with anencephaly—a disorder in which the brain is missing—also lack a nose. Without the nose the brain might suffer severely in its development.

Pasquale Graziadei, Biologist (1986)

58.
The Planets

The worlds of our solar system are widely different, but all share a common gravitational tie to the sun.

ISAAC ASIMOV

58.1 And God said, "Let there be lights in the firmament of the heavens to separate the day from the night; and let them be for signs and for seasons and for days and years, and let them be lights in the firmament of the heavens to give light upon the earth." And it was so. And God made the two great lights, the greater light to rule the day, and the lesser light to rule the night; he made the stars also.
The Bible **(circa 725 B.C.)**

58.2 Between this body [the earth] and the heavens there are suspended, in this aerial spirit, seven stars, separated by determinate spaces, which, on account of their motion, we call wandering.
Pliny the Elder (Gaius Plinius Secundus), Roman Naturalist/Historian (23–79)

58.3 The beauteous planet Venus, which invites to love, made all the orient laugh, outshining the light of the Fishes, which followed close behind her.
Dante Alighieri, Italian Poet (1265–1321)

58.4 The perfection of the heavenly spheres does not depend upon the order of their relative position as to whether one is higher than another.
Nicole Oresme, French Philosopher/Clergyman (1323–1382)

58.5 Also the earth is not spherical, as some have said, although it tends toward sphericity, for the shape of the universe is limited in its parts as well as its movement. . . . The movement which is more perfect than others is, therefore, circular, and the corporeal form which is the most perfect is the sphere.

Nicholas of Cusa, German Cardinal/Mathematician/Statesman (1401–1464)

58.6 First and above all lies the sphere of the fixed stars, containing itself and all things, for that very reason immovable; in truth, the frame of the universe, to which the motion and position of all other stars are referred. Of the moving bodies first comes Saturn, who completes his circuit in thirty years. After him, Jupiter, moving in a twelve-year revolution. Then Mars, who revolves biennially. Fourth in order an annual cycle takes place, in which we have said is contained the earth, with the lunar orbit as an epicycle. In the fifth place Venus is carried round in nine months. Then Mercury holds the sixth place, circulating in the space of eighty days. In the middle of all dwells the sun.

Nicholas Copernicus, Polish Astronomer (1473–1543)

58.7 But that which will excite the greatest astonishment by far, and which indeed especially moved me to call the attention of all astronomers and philosophers, is this: namely, that I have observed four planets, neither known nor observed by any one of the astronomers before my time, which have their orbits round a certain bright star [Jupiter], one of those previously known, like Venus or Mercury round the sun, and are sometimes in front of it, sometimes behind it, though they never depart from it beyond certain limits. All of which facts were discovered and observed a few days ago by the help of a telescope devised by me, through God's grace first enlightening my mind.

Galileo Galilei, Italian Astronomer/Physicist (1564–1642)

58.8 We did also at night see Jupiter and his girdle and satellites, very fine, with my twelve-foot glass, but could not Saturn, he being very dark.

Samuel Pepys, English Diarist (1633–1703)

58.9 We see it [Neptune] as Columbus saw America from the coast of Spain. Its movements have been felt, trembling along the far-reaching line of our analysis with a certainty hardly inferior to that of ocular demonstration.

Sir William Herschel, English Astronomer (1738–1822)

58.10 How can a man sit down and quietly pare his nails, while the earth goes gyrating ahead amid such a din of sphere music, whirling him along about her axis some twenty-four thousand miles between sun and sun, but mainly in a circle some two millions of miles actual progress? And then such a hurly-burly on the surface . . . . Can man do less than get up and shake himself?

Henry David Thoreau, American Writer/Naturalist (1817–1862)

58.11 What shall be the lot of attendant worlds that circle around such orbs; or of the earth as the Sun shall fade and cool?

E. Ledger, American Writer (1900)

58.12 The earth, formed out of the same debris of which the sun was born, is extraordinarily rich in iron—iron which once may have existed at the center of a star that exploded many billions of years ago.

Isaac Asimov, American Biochemist/Author (1965)

58.13 Eventually man has to get there [Mars] because we will never be satisfied with unmanned exploration.

Cyril Ponnamperuma, American Chemist (1985)

58.14 A hundred astronomers have left parts of their souls and their hopes in drawings showing the surface of Mars. A score of men have left their stamp in the major theories about life on the strange planet fourth from the sun. The names of ten thousand technicians and scientists rest now on a plaque standing a few feet above the soil of Mars, attached to a spacecraft sent there in 1976. Fifty writers have tried their pen out on Mars and things Martian; sixty movie directors have tried to grasp the magic and mystery. . . . I would like to show you how to fall in love with a planet.

Robert M. Powers, British Astronomer/Author (1986)

58.15 Venus does not have active volcanoes: two American scientists have reached this conclusion after reanalyzing data from the Pioneer Venus Orbiter spacecraft. They conclude that radio bursts detected by the spacecraft are not caused by lightning playing around active volcanoes, but are generated in the planet's ionosphere.

Nigel Henbest, Science Journalist (1986)

59. Probability

Religion considers the Universe deterministic

and science considers it probabilistic—

an important distinction.

ISAAC ASIMOV

59.1 How often things occur by mere chance which we dared not even hope for.

Terence (Publius Terentius Afer), Roman Dramatist (185 B.C.?–159 B.C.?)

59.2 Is it possible that a promiscuous jumble of printing letters should often fall into a method which should stamp on paper a coherent discourse?

John Locke, English Philosopher (1632–1704)

59.3 These duplicates in those parts of the body, without which a man might have very well subsisted, though not so well as with them, are a plain demonstration of an all-wise Contriver, as those more numerous copyings which are found among the vessels of the same body are evident demonstrations that they could not be the work of chance. This argument receives additional strength if we apply it to every animal and insect within our knowledge, as well as to those numberless living creatures that are objects too minute for a human eye: and if we consider how the several species in this whole world of life resemble one another in very many particulars, so far as is convenient for their respective states of existence, it is much more probable that a hundred millions of dice should be casually thrown a hundred millions of times in the same number than that the body of any single animal should be produced by the fortuitous concourse of matter.

Joseph Addison, English Essayist/Poet/Politician (1672–1719)

59.4 The laws of probability, so true in general, so fallacious in particular.
Edward Gibbon, English Historian (1737–1794)

59.5 There are things that are uncertain for us, things more or less probable, and we seek to compensate for the impossibility of knowing them by determining their different degrees of likelihood. So it is that we owe to the weakness of the human mind one of the most delicate and ingenious of mathematical theories, the science of chance or probability.
Pierre Simon Laplace, French Astronomer/Mathematician (1749–1827)

59.6 Chance favors the trained mind.
Louis Pasteur, French Chemist/Microbiologist (1822–1895)

59.7 A very small cause which escapes our notice determines a considerable effect that we cannot fail to see, and then we say that the effect is due to chance. If we knew exactly the laws of nature and the situation of the universe at the initial moment, we could predict exactly the situation of that same universe at a succeeding moment.
Jules Henri Poincaré, French Mathematician/Astronomer (1854–1912)

59.8 It is never possible to predict a physical occurrence with unlimited precision.
Max Planck, German Physicist (1858–1947)

59.9 Quantum mechanics is very impressive . . . but I am convinced that God does not play dice.
Albert Einstein, German/American Physicist (1879–1955)

59.10 It is easy without any profound logical analysis to perceive the difference between a succession of favorable deviations from the laws of chance, and on the other hand, the continuous and cumulative action of these laws.
Ronald Aylmer Fisher, English Geneticist (1890–1962)

59.11 Chance favors only those who know how to court her.
Charles-Jean-Henri Nicolle, French Physician/Bacteriologist (1932)

59.12 Heisenberg may have slept here.
American Bumper Sticker (1978)

60. Psychology and Psychiatry

Psychology marks the triumph of human evolution. How many other species would need a science of the mind?

ISAAC ASIMOV

60.1 It is not improbable that some of the presentations which come before the mind in sleep may even be causes of the actions cognate to each of them. For as when we are about to act [in waking hours], or are engaged in any course of action, or have already performed certain actions, we often find ourselves concerned with these actions, or performing them, in a vivid dream.
Aristotle, Greek Philosopher (384 B.C.–322 B.C.)

60.2 From the gods comes the saying "Know thyself."
Juvenal (Decimus Junius Juvenalis) Roman Poet/Satirist (60?–127?)

60.3 When I was a child, I spoke as a child, I understood as a child, I thought as a child: but when I became a man, I put away childish things.
The Bible **(circa 325 A.D.)**

60.4 Three sparks—pride, envy, and avarice—are those that have been kindled in all hearts.
Dante Alighieri, Italian Poet (1265–1321)

60.5 These fight not only with their hands, but with their heads, their breasts, and feet, rending each other piecemeal with their teeth. The good instructor said, "Now see, my son, the souls of those who were overcome with wrath."

Dante Alighieri, Italian Poet (1265–1321)

60.6 For men, to speak generally, are ungrateful, fickle, hypocritical, fearful of danger, and covetous of gain.

Niccolò Machiavelli, Italian Statesman/Political Philosopher (1469–1527)

60.7 This, as you know, is my opinion, that as the body when it tyrannizes over the mind ruins and destroys all its soundness, so in the same way when the mind becomes the tyrant, and not merely the true lord, it wastes and destroys the soundness of the body first, and then their common bond of union . . . and sins against prudence and charity.

Girolamo Fracastoro, Italian Physician/Poet (1478–1533)

60.8 Dreams are true interpreters of our inclinations; but there is art required to categorize and understand them.

Michel de Montaigne, French Essayist (1533–1592)

60.9 For behavior, men learn it, as they take diseases, one of another.

Francis Bacon, English Philosopher/Essayist/Statesman (1561–1626)

60.10 Men fear death as children fear to go in the dark; and as that natural fear in children is increased with tales, so is the other.

Francis Bacon, English Philosopher/Essayist/Statesman (1561–1626)

60.11 *Macbeth:* How does your patient, doctor?
Doctor: Not so sick, my lord,
As she is troubled with thick-coming fancies,
That keep her from her rest.
Macbeth: Cure her of that.
Canst thou not minister to a mind diseased,
Pluck from the memory a rooted sorrow,
Raze out the written troubles of the brain
And with some sweet oblivious antidote
Cleanse the stuff'd bosom of that perilous stuff
Which weighs upon the heart?
Doctor: Therein the patient
Must minister to himself.
Macbeth: Throw physic to the dogs; I'll none of it.

William Shakespeare, English Dramatist/Poet (1564–1616)

60.12 I'll change my state with any wretch
Thou canst from gaol of dunghill fetch.
My pain's past cure, another hell;
I may not in this torment dwell.
Now desperate I hate my life,

Lend me a halter or a knife!
All my griefs to this are jolly,
Naught so damned as melancholy.
Robert Burton, English Clergyman/Author (1577–1640)

60.13 So that in the nature of man, we find three principal causes of quarrel. First, competition; secondly, diffidence; thirdly, glory.
Thomas Hobbes, English Philosopher (1588–1679)

60.14 It may be in this case as it is with waters when their streams are stopped or dammed up, when they gett passage they flow with more violence, and make more noys and disturbance, then when they are suffered to rune quietly in their owne chanels. So wikednes being here more stopped by strict laws, and the same more nerly looked unto, so as it cannot rune in a comone road of liberty as it would, and is inclined, it searches every wher, and at last breaks out wher it getts vente.
William Bradford, American Governor (1590–1657)

60.15 The childhood shows the man
As morning shows the day.
John Milton, English Poet (1608–1674)

60.16 The humors of the body follow a uniform and regular course, which insensibly moves and turns our will. They circulate together and exercise in succession a secret power over us, so as to exert a considerable influence on our actions without our being aware of it.
François de la Rochefoucauld, French Writer/Moralist (1613–1680)

60.17 The state of man is changeableness, ennui, anxiety.
Blaise Pascal, French Mathematician/Philosopher/Author (1623–1662)

60.18 Our spirit is often led astray by its own delusions; it is even frightened by its own work, believes that it sees what it fears, and in the horror of night sees at last the objects which itself has produced.
François-Marie Arouet de Voltaire, French Author/Philosopher (1694–1778)

60.19 Men are like plants; the goodness and flavor of the fruit proceeds from the peculiar soil and exposition in which they grow. We are nothing but what we derive from the air we breathe, the climate we inhabit, the government we obey, the system of religion we profess, and the nature of our employment. Here you will find but few crimes; these have acquired as yet no root among us.
Hector St. John de Crevecoeur, American Farmer/French Consul to New York (1735–1813)

60.20 By what astonishing power does it come to pass that Man can so thoroughly imbibe the instinct and adopt the ferocity of the tiger, and yet be so indifferent in his faculties and organs? The tiger sheds no blood but when impelled to it by the stings of hunger; had Nature taught him to eat grass, he would not be the tiger. But Man, who eats no man, yet kills Man and takes a singular pleasure in shedding his blood.

Hector St. John de Crevecoeur, American Farmer/French Consul to New York (1735–1813)

60.21 The mind of a young man (his gallery I mean) is often furnished different ways. According to the scenes he is placed in, so are his pictures. They disappear, and he gets a new set in a moment. But as he grows up, he gets some substantial pieces which he always preserves, although he may alter his smaller paintings in a moment.

James Boswell, Scottish Lawyer/Author (1740–1795)

60.22 I complained to Mr. Johnson that I was much afflicted with melancholy, which was hereditary in our family. He said that he himself had been greatly distressed with it, and for that reason had been obliged to fly from study and meditation to the dissipating variety of life. He advised me to have constant occupation of mind, to take a great deal of exercise, and to live moderately; especially to shun drinking at night. "Melancholy people," said he, "are apt to fly to intemperance, which gives a momentary relief but sinks the soul much lower in misery." He observed that laboring men who work much and live sparingly are seldom or never troubled with low spirits.

James Boswell, Scottish Lawyer/Author (1740–1795)

60.23 In general, we receive impressions only in consequence of motion, and we might establish it as an axiom *that without motion there is no sensation.*

Antoine Lavoisier, French Chemist (1743–1794)

60.24 The naturalists, you know, distribute the history of nature into three kingdoms or departments: zoology, botany, mineralogy. Ideology, or mind, however, occupies so much space in the field of science, that we might perhaps erect it into a fourth kingdom or department. But inasmuch as it makes a part of the animal construction only, it would be more proper to subdivide zoology into physical and moral.

Thomas Jefferson, American President/Author (1743–1826)

60.25 Shakespeare was pursuing two Methods at once; and besides the Psychological Method, he had also to attend to the Poetical. (Note) we beg pardon for the use of this insolent verbum: but it is one of which our Language stands in great need. We have no single term to express the Philosophy of the Human Mind.

Samuel Taylor Coleridge, English Poet/Critic (1772–1834)

60.26 The most general survey shows us that the two foes of human happiness are pain and boredom.

Arthur Schopenhauer, German Philosopher (1788–1860)

60.27 I would paint what has not been unhappily called the psychological character.

Benjamin Disraeli, English Statesman/Politician/Writer (1804–1881)

60.28 Whether the minds of men and women are or are not alike, are obviously psychological questions.

Herbert Spencer, English Philosopher/Psychologist (1820–1903)

60.29 Psychology is a part of the science of life or biology. . . . As the physiologist inquires into the way in which the so-called "functions" of the body are performed, so the psychologist studies the so-called "faculties" of the mind.

Thomas Henry Huxley, English Biologist/Philosopher (1825–1895)

60.30 It is of great psychological interest to follow up the gradual development of civilization and the influence exerted by sexual life upon habits and morality. The gratification of the sexual instinct seems to be the primary motive in man as well as in beast.

Richard von Krafft-Ebing, German Psychiatrist/Neurologist (1840–1902)

60.31 Consciousness . . . does not appear to itself chopped up in bits. Such words as "chain" or "train" do not describe it fitly as it presents itself in the first instance. It is nothing jointed; it flows. A "river" or a "stream" is the metaphor by which it is most naturally described. In talking of it hereafter, let us call it the stream of thought, of consciousness, or of subjective life.

William James, American Psychologist/Philosopher (1842–1910)

60.32 The mind, in short, works on the data it receives very much as a sculptor works on his block of stone. In a sense the statue stood there from eternity. But there were a thousand different ones beside it, and the sculptor alone is to thank for having extricated this one from the rest. Just so with the world of each of us, howsoever different our several views of it may be, all lay embedded in the primordial chaos of sensations, which gave the mere *matter* to the thought of all of us indifferently.

William James, American Psychologist/Philosopher (1842–1910)

60.33 Every perception is an acquired perception.

William James, American Psychologist/Philosopher (1842–1910)

60.34 Not everything is an idea. Otherwise psychology would contain all the sciences within it or at least it would be the highest judge over all the sciences. Otherwise psychology would rule over logic and mathematics. But nothing would be a greater misunderstanding of mathematics than its subordination to psychology.

Gottlob Frege, German Mathematician/Logician (1848–1925)

60.35 Only science, exact science about human nature itself, and the most sincere approach to it by the aid of the omnipotent scientific method, will deliver man from his present gloom and will purge him from his contemporary share in the sphere of interhuman relations.

Ivan Petrovich Pavlov, Russian Physiologist (1849–1936)

60.36 The genesis of mathematical creation is a problem which should intensely interest the psychologist.

Jules Henri Poincaré, French Mathematician/Astronomer (1854–1912)

60.37 The symptoms of neurosis, as we have learnt, are essentially substitute gratifications for unfulfilled sexual wishes.

Sigmund Freud, Austrian Psychiatrist/Psychoanalyst (1856–1939)

60.38 Every dream reveals a psychological structure, full of significance.

Sigmund Freud, Austrian Psychiatrist/Psychoanalyst (1856–1939)

60.39 A neurosis is the result of a conflict between the ego and the id; the person is at war with himself. A psychosis is the outcome of a similar disturbance in the relation between the ego and the outside world.

Sigmund Freud, Austrian Psychiatrist/Psychoanalyst (1856–1939)

60.40 Nevertheless, his [Dostoevsky's] personality retained sadistic traits in plenty, which show themselves in his irritability, his love of tormenting, and his intolerance even towards people he loved, and which appear also in the way in which, as an author, he treats his readers. Thus in little things he was a sadist towards others, and in bigger things a sadist towards himself, in fact a masochist—that is to say the mildest, kindliest, most helpful person possible.

Sigmund Freud, Austrian Psychiatrist/Psychoanalyst (1856-1939)

60.41 In psychoanalytic treatment nothing happens but an exchange of words between the patient and the physician.

Sigmund Freud, Austrian Psychiatrist/Psychoanalyst (1856–1939)

60.42 Wit is the best safety valve modern man has evolved; the more civilization, the more repression, the more the need there is for wit.

Sigmund Freud, Austrian Psychiatrist/Psychoanalyst (1856–1939)

60.43 In the training and in the exercise of medicine a remoteness abides between the field of neurology and that of mental health, psychiatry. It is sometimes blamed to prejudice on the part of the one side or the other. It is both more grave and less grave than that. It has a reasonable basis. It is rooted in the energy-mind problem. Physiology has not enough to offer about the brain in relation to the mind to lend the psychiatrist much help.

Sir Charles Scott Sherrington, English Physiologist (1857–1952)

60.44 The psyche is distinctly more complicated and inaccessible than the body. It is, so to speak, the half of the world which comes into existence only when we become conscious of it. For that reason the psyche is not only a personal but a world problem, and the psychiatrist has to deal with an entire world.

Carl Gustav Jung, Swiss Psychiatrist/Psychologist (1875–1961)

60.45 There are as many archetypes as there are typical situations in life.

Carl Gustav Jung, Swiss Psychiatrist/Psychologist (1875–1961)

60.46 The dream is a little hidden door in the innermost and most secret recesses of the psyche, opening into the cosmic night which was psyche long before there was any ego-consciousness, and which will remain psyche no matter how far our ego-consciousness may extend.

Carl Gustav Jung, Swiss Psychiatrist/Psychologist (1875–1961)

60.47 Personality is the supreme realization of the innate idiosyncracy of a living being. It is an act of high courage flung in the face of life.

Carl Gustav Jung, Swiss Psychiatrist/Psychologist (1875–1961)

60.48 The genius which runs to madness is no longer genius.

Otto Weininger, German Psychologist (1880–1903)

60.49 We are too ready to accept others and ourselves as we are and to assume that we are incapable of change. We forget the idea of growth, or we do not take it seriously. There is no good reason why we should not develop and change until the last day we live. Psychoanalysis is one of the most powerful means of helping us to realize this aim.

Karen Horney, American Psychoanalyst (1885–1952)

60.50 [Freud] would often say three things were impossible to fulfill completely: healing, educating, governing. He limited his goals in analytic treatment to bringing a patient to the point where he could *work* for a living, and learn to *love*.

Theodore Reik, Austrian/American Psychoanalyst (1888–1969)

60.51 A neurotic person can be most simply descibed as someone who, while he was growing up, learned ways of behaving that are self-defeating in his society.

Margaret Mead, American Anthropologist (1901–1978)

60.52 If it is psychically or physically in need of emotional peace, physical regeneration, and release from tension or stress, then the instinctive response will be to choose the darker colors. If the organism needs to dissipate energy by outgoing activity or in mental creativeness, then the instinctive response will be for the brighter colors.

Max Lüscher, Swiss Psychologist (1948)

60.53 When colors have been accurately chosen for their direct association with psychological and physiological needs . . . a preference for one color and a dislike for another means something definite and reflects an existing state of mind, or glandular balance, or both. To see how this can be so, why this relationship is so universal and why it exists independent of race, sex, or social environment, it is necessary to look back at man's long exposure to the colors of Nature.

Max Lüscher, Swiss Psychologist (1948)

60.54 The misconceptions and distortions, the falsifications and misrepresentations to which psychoanalysis was subjected in its popularizations, threaten to transform the magnificent house that Freud built into a stable similar to that of King Augeus.

Theodore Reik, Austrian/American Psychoanalyst (1950)

60.55 If Freud is not the Pasteur of mental illness, he is at least its Hippocrates.

Isaac Asimov, American Biochemist/Author (1965)

60.56 We live, after all in the age of "behavioral science," not of "the science of mind."... No sane person has ever doubted that behavior provides much of the evidence for this study—all of the evidence, if we interpret "behavioral science" suggests a not-so-subtle shift of emphasis toward the evidence itself and away from the deeper underlying principles and abstract mental structures that might be illuminated by the evidence of behavior. It is as if natural science were to be designated "the science of meter readings."

Noam Chomsky, American Linguist/Philosopher (1968)

60.57 The Gay Liberation movement is the best therapy the homosexual has had in years.

Lawrence Leshan, American Psychologist (1971)

60.58 Many peptic ulcers and psychosomatic ailments are poems struggling to be born.

Jack Leedy, American Psychologist/Poetry Therapist (1971)

60.59 I felt the experience of being out in the wilderness would add another dimension to the psychotherapy experience. . . . In the woods there is a whole new set of rules and obligations: this process of deculturation helps people to relate to each other openly and honestly.

Arthur Kovacs, American Psychologist (1971)

60.60 We'd rather take the risk of AWOL patients than lock them up. Freedom is better for the patient's morale. Locked doors have a very depressing effect on the patient.

Andrew K. Bernath, American Psychiatrist/Hospital Director (1971)

60.61 Psychology . . . a sort of PacMan of the social sciences, that is, busy, ambitious, imperial, eager to gobble up all the subject matter in sight.

Joseph Adelson, American Psychologist/Author (1980)

60.62 In a way, all psychotherapy is reprogramming. You aren't curing a disorder; you're adding something to the history of the individual. If what you add is powerful enough, it will certainly lead to different behavior.

B.F. Skinner, American Psychologist (1984)

60.63 Incredible shame is associated with mental illness. People will confide the most intimate details of their love life before they'll mention a relative who has had a serious mental breakdown. But the brain is just another organ. It's just a machine, and a machine can go wrong.

Candace Pert, American Neuroscientist (1984)

60.64 The highest intelligence on the planet probably exists in a sperm whale, who has a ten-thousand-gram brain, six times larger than ours. I'm convinced that intelligence is a function of absolute brain size.

John Lilly, American Neurophysiologist/Spiritualist (1984)

60.65 In the province of the mind, what is believed to be true is true, or becomes true, within limits to be learned by experience and experiment. These limits are further beliefs to be transcended. In the province of the mind there are no limits.

John Lilly, American Neurophysiologist/Spiritualist (1984)

60.66 My hopes for LSD were absolutely concentrated on the psychiatric field. From my own experience, I realized that LSD could be a useful agent in psychoanalysis and psychotherapy, that patients could leave their everyday, ordinary reality with it by getting out of their problems and into another sphere of consciousness. I also thought it could be important for brain research.

Albert Hofmann, Swiss Chemist (1984)

60.67 Torturing became a job. If the officers ordered you to beat, you beat. If they ordered you to stop, you stopped. You never thought you could do otherwise.

Michaelis Petrou, Greek Torturer (1986)

60.68 For me, the most difficult thing was to realize how cruel other men can be.

Giorgio Solimano, Chilean Physician/Educator/Torture Victim (1986)

60.69 They [torture victims] may feel better, they may have families and function at a high level, but they will always remember.

Glenn Randall, American Internist (1986)

60.70 One person might go out and rob banks or beat up old ladies in the park, while someone with the same propensity [for aggression] in a different environment might run a multinational corporation.

June Reinisch, American Director of the Kinsey Institute (1986)

60.71 It appears that [physical] fitness may be a valuable prophylactic for dealing with stress.

David S. Holmes, American Psychologist (1986)

60.72 I've a fundamental rule that for anybody who's in serious trouble, it's got to be either sex or death. If it's death, it's suicide or homicide. And if it's sex, it's animal, mineral, or vegetable; man, woman, or child.

John Money, American Psychoendocrinologist (1986)

60.73 Consciousness isn't like a light switch that goes on or off. It's more like a dimmer on a light.

Nathan Cope, American Anthropologist (1986)

60.74 I have nothing to offer except a way of looking at things.

Erik H. Erikson, American Psychologist (1986)

61. Radiation

Radiation, unlike smoking, drinking, and overeating, gives no pleasure, so the possible victims object.

ISAAC ASIMOV

61.1 Radiant energy, which at the beginning [of the universe] played a predominant role in the evolutionary process, gradually lost its importance and by the end of the thirty-millionth year yielded its priority in favor of ordinary atomic matter.

Werner Karl Heisenberg, German Physicist (1901–1976)

61.2 Little Willie, full of glee,
Put radium in Grandma's tea.
Now he thinks it quite a lark
To see her shining in the dark.

Dorothy Rickard, American Writer (1953)

61.3 The fact that the general incidence of leukemia has doubled in the last two decades may be due, partly, to the increasing use of x-rays for numerous purposes. The incidence of leukemia in doctors, who are likely to be so exposed, is twice that of the general public. In radiologists . . . the incidence is ten times greater.

Isaac Asimov, American Biochemist/Author (1965)

61.4 Nuclear power plants are always emitting radiation. It isn't a continuous thing, it's sporadic. It's a part of the routine operation especially when they begin the operation or shut down. And you never know how much is being emitted. That was in the paper. . . . Met. Ed. admits it and the NRC admits it. There's a certain amount of emissions coming from a nuclear plant.

Jane Lee, American Dairy Farmer, Three Mile Island Area (1980)

61.5 There was no trouble with these calves at the dairy farms, but when the dairy farmers sold them to their neighbors, there were problems like multiple fractures, blindness, chronic arthritis, anemia, and what looked like starvation and deficiency. . . . We've seen this problem for about two-and-a-half years before the accident. The last case was in June, three months after the accident. Since the plant has been closed, we haven't run into this problem.

Dr. Robert Weber, American Veterinarian, Three Mile Island Area (1980)

61.6 It could be that we're losing some of them [minerals and other nutrients in the soil] because of TMI. I've seen it when water runs out a rainspout across the lawn. For perhaps twenty feet out, a strip of grass about two-and-a-half feet wide is all brown. I never saw it before this spring. But even under the barbed wire or the wire fences, there's a strip, say three inches wide, where the grass is dead. And there's no explanation for what happened to some of the trees I've seen.

Vance Fisher, American Farmer, Three Mile Island Area (1980)

61.7 We found no animals with any detectable contamination. The Pennsylvania Department of Agriculture has visited each of the farms that claimed to have problems, and a report on that was supposed to be out this week. . . . But the information I have is that there's nothing abnormal at all in the Three Mile Island area, according to the Pennsylvania Department of Agriculture.

Thomas M. Gerusky, American Environmental Administrator (1980)

61.8 The only health impact of the Three Mile Island accident that can be identified with certainty is mental stress to those living in the vicinity of the plant.

Arthur C. Upton, American Physician (1980)

61.9 Hell no, we won't glow.

American Antinuclear Slogan (June 12, 1982)

61.10 Until we understand how microwave radiation interacts with living systems, the public will continue to be suspicious and the microwave problem will continue to be with us.

Louis Slesin, American Editor (*Microwave News*) (1986)

61.11 In the schools [in Kiev] Geiger counters are no longer a part of the civil defense demonstration kit, but a necessity of everyday housekeeping. . . . In at least one math class, the basic lesson was: "The smallest atom bomb equals three Chernobyls."

Felicity Barringer, American Journalist (1986)

62. Reproduction

To prevent overpopulation there must be more deaths or fewer births. Refuse the latter; there will be the former.

ISAAC ASIMOV

62.1 So the Lord God caused a deep sleep to fall upon the man, and while he slept took one of his ribs and closed up its place with flesh; and the rib which the Lord God had taken from the man he made into a woman and brought her to the man.

The Bible **(circa 725 B.C.)**

62.2 Be fruitful and multiply, and then fill the earth and subdue it.

The Bible **(circa 725 B.C.)**

62.3 I will not give any woman the instrument to procure abortion.

Hippocrates, Greek Physician (460 B.C.–377 B.C.)

62.4 Men, in most instances, continue to be sexually functional until they are sixty years old, and, if that limit is exceeded, then until seventy years; and men have been actually known to bring forth progeny at seventy years of age.

Aristotle, Greek Philosopher (384 B.C.–322 B.C.)

62.5 Animals generally seem naturally disposed to intercourse at about the same period of the year, and that is when winter changes to summer. . . . In the human species, the male experiences more sexual excitement in winter, and the female in summer.

Aristotle, Greek Philosopher (384 B.C.–322 B.C.)

62.6 All living creatures, whether they swim, or walk, or fly, and whether they come into the world with the form of an animal or of an egg, are engendered in the same way.

Aristotle, Greek Philosopher (384 B.C.–322 B.C.)

62.7 Every animal is sad after coitus except the human female and the rooster.

Claudius Galen, Greek Physician/Scholar (130–200)

62.8 It is as natural to die as to be born; and to a little infant, perhaps, the one is as painful as the other.

Francis Bacon, English Philosopher/Essayist/Statesman (1561–1626)

62.9 Macduff was from his mother's womb
Untimely ripped.

William Shakespeare, English Dramatist/Poet (1564–1616)

62.10 A weird happening has occurred in the case of a lansquenet named Daniel Brughammer, of the squadron of Captain Burkhard Laymann Zu Liebenau,of the honorable Madrucci Regiment in Piadena, in Italy. When the same was on the point of going to bed one night he complained to his wife, to whom he had been married by the Church seven years ago, that he had great pains in his belly and felt something stirring therein. An hour thereafter he gave birth to a child, a girl. When his wife was made aware of this, she notified the occurrence at once. Thereupon he was examined and questioned. . . . He confessed on the spot that he was half man and half woman and that for more than seven years he had served as a soldier in Hungary and the Netherlands. . . . When he was born he was christened as a boy. . . . He also stated that while in the Netherlands he only slept once with a Spaniard, and he became pregnant therefrom. This, however, he kept a secret unto himself and also from his wife, with whom he had for seven years lived in wedlock, but he had never been able to get her with child. . . . The aforesaid soldier is able to suckle the child with his right breast only and not at all on the left side, where he is a man. He has also the natural organs of a man for passing water. Both are well, the child is beautiful, and many towns have already wished to adopt it, which, however, has not as yet been arranged. All this has been set down and described by notaries. It is considered in Italy to be a great miracle.

Anonymous, Business Agent for the Fugger Banking Family (1589)

62.11 I would be content that we might procreate like trees, without conjunction, or that there were any way to perpetuate the World without this trivial and vulgar way of coition.

Thomas Browne, English Writer/Antiquarian (1605–1682)

62.12 And therefore though Adam was framed without this part (a navel), as having no other womb than that of his proper principles, yet was not his posterity without the same: for the seminality of his fabric contained the power thereof; and was endued with the science of those parts whose predestinations upon succession it did accomplish.

Thomas Browne, English Writer/Antiquarian (1605–1682)

62.13 By the act of generation nothing more is done than to ferment the sperm of the female by the sperm of the male that it may thereby become fit nourishment for the Embryo: for the nourishment of all animals is prepared by ferment and the ferment is taken from animals of the same kind, and makes the nourishment subtile and spiritual. In adult animals the nourishment is fermented by the choler and pancreatic juice both which come from the blood. The Embryo not being able to ferment its own nourishment which comes from the mother's blood has it fermented by the sperm which comes from the father's blood, and by this nourishment it swells, drops off from the Ovarium and begins to grow with a life distinct from that of the mother.

Isaac Newton, English Physicist/Mathematician (1642–1727)

62.14 Population, when unchecked, increases in geometrical ratio. Subsistence increases only in an arithmetical ratio.

Thomas Robert Malthus, English Clergyman/Economist (1766–1834)

62.15 That a woman should at present be almost driven from society, for an offence [illegitimate childbearing], which men commit nearly with impunity, seems to be undoubtedly a breach of natural justice. . . . What at first might be dictated by state necessity, is now supported by female delicacy; and operates with the greatest force on that part of society, where, if the original intention of the custom were preserved, there is the least real occasion for it.

Thomas Robert Malthus, English Clergyman/Economist (1766–1834)

62.16 The propagation of the human race is not left to mere accident or the caprices of the individual, but is guaranteed by the hidden laws of nature which are enforced by a mighty, irresistible impulse.

Richard von Krafft-Ebing, German Psychiatrist/Neurologist (1840–1902)

62.17 Altering a gene in the gene line to produce improved offspring is likely to be very difficult because of the danger of unwanted side effects. It would also raise obvious ethical problems.

Francis Crick, British Biophysicist/Geneticist (1916–)

62.18 If the earth's population continues to double every 50 years (as it is now doing) then by 2550 A.D. it will have increased 3,000-fold. . . . by 2800 A.D., it would reach 630,000 billion! Our planet would have standing room only, for there would be only two-and-a-half square feet per person on the entire land surface, including Greenland and Antarctica. In fact, if the human species could be imagined as continuing to multiply further at the same rate, by 4200 A.D. the total mass of human tissue would be equal to the mass of the earth.

Isaac Asimov, American Biochemist/Author (1965)

62.19 The battle to feed all humanity is over. In the 1970s the world will undergo famines—hundreds of millions of people are going to starve to death.

Dr. Paul Ehrlich, American Biologist/Ecologist (1968)

62.20 Modern times breed modern phobias. Until the present age, for example, it has been impossible for any woman to suffer crippling fear of artificial insemination.

Russell Baker, American Writer (1986)

62.21 Today, a dozen European countries have nearly steady or falling population levels, and the fertility rate for the entire developed world has dropped below the replacement level.

The latest explosion, in which the population rose from 800 million in 1750 to over 4.5 billion today, already seems to be abating. Since 1973 the world's growth rate has once again slowed (from 2 percent to 1.7 percent) as fertility rates have dropped sharply in Asia and Latin America.

John Tierney, American Science Journalist (1986)

62.22 Freedom to breed will bring ruin to all.

Garrett Hardiny, American Ecologist (1986)

62.23 For example, to whom do such embryos [cryopreserved human embryos] "belong"—to living parents, to the estate of deceased parents, to the storage facility that maintains them, to the state?

Clifford Grobstein, American Biologist/Educator (1986)

62.24 When a low-ranking animal is attacked in early pregnancy, it could cause her to abort. From her point of view, ovulating or starting a pregnancy when others are in heat just isn't worth it. Rather than lose a valuable year of her reproductive career carrying and nursing an infant that might well get harassed to death, she defers fertility until a less competitive time of year . . . the hormones her body releases under stress make her reproductive system shut down.

Duncan Maxwell Anderson, American Writer (1986)

62.25 Although the diaphragm had been in existence for ages, the pill's incredible value was that you put it in your mouth, not in your vagina. So it wasn't sex. It was so completely de-eroticized, it was acceptable.

John Money, American Psychoendocrinologist (1986)

62.26 All of us should be greatly concerned by the marketing of human reproduction, not a matter to be governed by dollars and by power. It could easily degenerate into a matter of rich people's paying poor people to go through the discomforts and the potential risks of pregnancy, much as wealthy people in the Civil War paid to have impoverished men serve in the Army in their stead.

Arthur W. Feinburg, American Medical Educator (1986)

62.27 Most critical is whether the entire process of surrogate motherhood leads to a dehumanization of the most human activity we know, the reproduction of our own species.

Arthur W. Feinburg, American Medical Educator (1986)

63.
Research

"Research" means "to search again."

Why not? Sometimes, a new interpretation

emerges that is of vast importance.

ISAAC ASIMOV

63.1 Nothing has such power to broaden the mind as the ability to investigate systematically and truly all that comes under thy observation in life.
Marcus Aurelius, Roman Emperor/Philosopher (121–180)

63.2 Study is like the heaven's glorious sun,
That will not be deep-search'd with saucy looks:
Small have continual plodders ever won,
Save base authority from others' books.
William Shakespeare, English Dramatist/Poet (1564–1616)

63.3 Attempt the end and never stand to doubt;
Nothing so hard but search will find it out.
Robert Herrick, English Poet (1591–1674)

63.4 Far must thy researches go
Wouldst thou learn the world to know;
Thou must tempt the dark abyss
Wouldst thou prove what *Being* is;
Naught but firmness gains the prize,
Naught but fullness makes us wise,
Buried deep truth ever lies.
Friedrich von Schiller, German Poet/Dramatist (1759–1805)

63.5 The mysterious and unsolved problem of how things came to be does not enter the empirical province of objective research, which is confined to a description of things as they are.

Alexander von Humboldt, German Naturalist/Statesman (1769–1835)

63.6 Lost in a gloom of uninspired research.

William Wordsworth, English Poet (1770–1850)

63.7 The investigator should have a robust faith—and yet not believe.

Claude Bernard, French Physiologist (1813–1878)

63.8 Don't spare; don't drudge.

Benjamin Jowett, English Scientist (1817–1893)

63.9 Endow the already established with money. Endow the woman who shows genius with time.

Maria Mitchell, American Astronomer (1818–1889)

63.10 There are not many joys in human life equal to the joy of the sudden birth of a generalization illuminating the mind after a long period of patient research.

Prince Pëtr Alekseevich Kropotkin, Russian Anarchist/Geologist (1842–1921)

63.11 Scientific discovery and scientific knowledge have been achieved only by those who have gone in pursuit of them without any practical purpose whatsoever in view.

Max Planck, German Physicist (1858–1947)

63.12 Research is fundamentally a state of mind involving continual re-examination of doctrines and axioms upon which current thought and action are based. It is, therefore, critical of existing practices.

Theobald Smith, American Pathologist (1859–1934)

63.13 The joy of research must be found in doing, since every other harvest is uncertain.

Theobald Smith, American Pathologist (1859–1934)

63.14 Those hateful persons called Original Researchers.

J.M. (James Matthew) Barrie, British Dramatist/Novelist (1860–1937)

63.15 War should mean research.

George Ellery Hale, American Astronomer (Former Director, National Research Council) (1868–1938)

63.16 The investigator may be made to dwell in a garret, he may be forced to live on crusts and wear dilapidated clothes, he may be deprived of social recognition, but if he has time, he can steadfastly devote himself to research.

Take away his free time and he is utterly destroyed as a contributor to knowledge.

Walter B. Cannon, American Physiologist (1871–1945)

63.17 The final results [of his work on the theory of relativity] appear almost simple; any intelligent undergraduate can understand them without much trouble. But the years of searching in the dark for a truth that one feels, but cannot express; the intense effort and the alternations of confidence and misgiving, until one breaks through to clarity and understanding, are only known to him who has himself experienced them.

Albert Einstein, German/American Physicist (1879–1955)

63.18 Committees are dangerous things that need most careful watching. I believe that a research committee can do one useful thing and one only. It can find the workers best fitted to attack a particular problem, bring them together, give them the facilities they need, and leave them to get on with the work. It can review progress from time to time, and make adjustments; but if it tries to do more, it will do harm.

W.W.C. Topley, English Physician/Medical Educator (1886–1944)

63.19 If we want an answer from nature, we must put our questions in acts, not words, and the acts may take us to curious places. Some questions were answered in the laboratory, others in mines, others in a hospital where a surgeon pushed tubes in my arteries to get blood samples, others on top of Pike's Peak in the Rocky Mountains, or in a diving dress on the bottom of the sea. That is one of the things I like about scientific research. You never know where it will take you next.

John Burdon Sanderson Haldane, English Geneticist (1892–1964)

63.20 Investigation may be likened to the long months of pregnancy, and solving a problem to the day of birth. To investigate a problem is, indeed, to solve it.

Mao Tse Tung, Chinese Political Leader (1893–1976)

63.21 I do not think we can impose limits on research. Through hundreds of thousands of years, man's intellectual curiosity has been essential to all the gains we have made. Although in recent times we have progressed from chance and hit-or-miss methods to consciously directed research, we still cannot know in advance what the results may be. It would be regressive and dangerous to trammel the free search for new forms of truth.

Margaret Mead, American Anthropologist (1901–1978)

63.22 Basic research is what I am doing when I don't know what I am doing.

Wernher von Braun, German/American Rocket Engineer (1912–1977)

63.23 Most of the knowledge and much of the genius of the research worker lie behind his selection of what is worth observing. It is a crucial choice, often determining the success or failure of months of work, often differentiating the brilliant discoverer from the . . . plodder.

Alan Gregg, English Scientist/Author (1941)

63.24 I became expert at dissecting crayfish. At one point I had a crayfish claw mounted on an apparatus in such a way that I could operate the individual nerves. I could get the several-jointed claw to reach down and pick up a pencil and wave it around. I am not sure that what I was doing had much scientific value, although I did learn which nerve fiber had to be excited to inhibit the effects of another fiber so that the claw would open. And it did get me interested in robotic instrumentation, something that I have now returned to. I am trying to build better micromanipulators for surgery and the like.

Marvin Minsky, American Computer Scientist (1982)

63.25 It's hard to imagine anything more difficult to study than human sexuality, on every level from the technical to the political. One has only to picture monitoring orgasm in the lab to begin to grasp the challenge of developing testing techniques that are thorough and precise, yet respectful.

Winnifred Gallagher, American Science Journalist (1986)

63.26 As a different, but perhaps more common, strategy for the suppression of novelty, we may admit the threatening object to our midst, but provide an enveloping mantle of ordinary garb. . . .

This kind of cover-up, so often amusing in our daily lives, can be quite dangerous in science, for nothing can stifle originality more effectively than an ordinary mantle placed fully and securely over an extraordinary thing.

Stephen Jay Gould, American Biologist/Author (1986)

64.
Science

Any increase in knowledge anywhere helps pave the way for an increase in knowledge everywhere.

ISAAC ASIMOV

64.1 Science is the father of knowledge, but opinion breeds ignorance.
Hippocrates, Greek Physician (460 B.C.?–377 B.C.?)

64.2 To be acceptable as scientific knowledge a truth must be a deduction from other truths.
Aristotle, Greek Philosopher (384 B.C.–322 B.C.)

64.3 Shun no toil to make yourself remarkable by some talent or other; yet do not devote yourself to one branch exclusively. Strive to get clear notions about all. Give up no science entirely; for science is but one.
Lucius Annaeus Seneca (the Younger), Roman Philosopher/Statesman/ Dramatist (4 B.C.?–65 A.D.)

64.4 Keep that which is committed to thy trust, avoiding profane and vain babblings, and oppositions of science falsely so called.
***The Bible* (circa 325 A.D.)**

64.5 All sciences are connected; they lend each other material aid as parts of one great whole, each doing its own work, not for itself alone, but for the other parts; as the eye guides the body and the foot sustains it and leads it from place to place.
Roger Bacon, English Philosopher/Clergyman (1220–1292)

64.6 Science without conscience is the death of the soul.
François Rabelais, French Monk/Physician/Satirist (1494?–1553?)

64.7 No tempest can consume Science.
John Florio, Italian/English Author (1533?–1625)

64.8 The strength of all sciences, which consisteth in their harmony, each supporting the other, is as the strength of the old man's fagot in the band; for were it not better for a man in a fair room to set up one great light, or branching candlestick of lights, than to go about with a small watch-candle into every corner?
Francis Bacon, English Philosopher/Essayist/Statesman (1561–1626)

64.9 Science [is the] knowledge of the truth of Propositions, and how things are called.
Thomas Hobbes, English Philosopher (1588–1679)

64.10 Much science, much sorrow.
John Clarke, English/American Clergyman (1609–1676)

64.11 Is it not evident, in these last hundred years, that more errors of the school have been detected, more useful experiments in philosophy have been made, more noble secrets in optics, medicine, anatomy, astronomy, discovered, than in all those credulous and doting ages, from Aristotle to us? So true it is, that nothing spreads more fast than science, when rightly and generally cultivated.
John Dryden, English Poet/Critic/Dramatist (1631–1700)

64.12 Old sciences are unraveled like old stockings, by beginning at the foot.
Jonathan Swift, Irish Satirist/Clergyman (1667–1745)

64.13 Science when well digested is nothing but good sense and reason.
Stanislaw I Leszczyński, Polish King (1677–1766)

64.14 Trace science then, with modesty thy guide;
First strip off all her equipage of pride;
Deduct what is but vanity, or dress,
Or learning's luxury, or idleness;
Or tricks to show the stretch of human brain,
Mere curious pleasure, or ingenious pain;
Expunge the whole, or log th' excrescent parts
Of all our vices have created arts;
Then see how little the remaining sum
Which serv'd the past, and must the times to come.
Alexander Pope, English Poet/Satirist (1688–1744)

64.15 Life is not the object of Science: we see a little, very little; And what is beyond we can only conjecture.
Samuel Johnson, English Lexicographer/Poet/Critic (1709–1784)

64.16 We do not enlarge but disfigure sciences, if we allow them to trespass on one another's territory.

Immanuel Kant, German Philosopher (1724–1804)

64.17 Nothing tends so much to the corruption of science as to suffer it to stagnate.

Edmund Burke, English Statesman/Author (1729–1797)

64.18 They may say what they like; everything is organized matter. The tree is the first link of the chain; man is the last. Men are young; the earth is old. Vegetable and animal chemistry are still in their infancy. Electricity, galvanism—what discoveries in a few years!

Napoleon I (Napoleon Bonaparte), French Emperor/General (1769–1821)

64.19 From Avarice thus, from Luxury and War
Sprang heavenly Science;
and from Science Freedom.

Samuel Taylor Coleridge, English Poet/Critic (1772–1834)

64.20 O star-eyed Science, hast thou wandered there,
To waft us home the message of despair?

Thomas Campbell, English Poet (1777–1844)

64.21 The origin of all science is the *desire to know causes;* and the origin of all false science and imposture is in the desire to accept false causes rather than none; or, which is the same thing, in the unwillingness to acknowledge our own ignorance.

William Hazlitt, English Critic/Essayist (1778–1830)

64.22 Who shall assign a limit to the discoveries of future ages? Who can prescribe to science her boundaries, or restrain the active and insatiable curiosity of man within the circle of his present acquirements?

Thomas Chalmers, Scottish Theologian/Author (1780–1847)

64.23 Knowledge is not happiness, and science
But an exchange of ignorance for that
Which is another kind of ignorance.

Lord Byron (George Gordon), English Poet/Dramatist (1788–1824)

64.24 The estimate we form of the intellectual capacity of our race, is founded on an examination of those productions which have resulted from the loftiest flights of individual genius, or from the accumulated labors of generations of men, by whose long-continued exertions a body of science has been raised up, surpassing in its extent the creative powers of any individual, and demanding for its development a length of time, to which no single life extends.

Charles Babbage, English Inventor/Mathematician (1792–1871)

64.25 Experience is the mother of science.
Henry George Bohn, English Publisher (1796–1884)

64.26 To understand a science it is necessary to know its history.
Auguste Comte, French Philosopher (1798–1857)

64.27 Men love to wonder, and that is the seed of our science.
Ralph Waldo Emerson, American Essayist/Philosopher/Poet (1803–1882)

64.28 Every great scientific truth goes through three states: first, people say it conflicts with the Bible; next, they say it has been discovered before; lastly, they say they always believed it.
Jean Louis Agassiz, Swiss/American Naturalist/Geologist (1807–1873)

64.29 Science is a first-rate piece of furniture for a man's upper chamber, if he has common sense on the ground floor.
Oliver Wendell Holmes, American Physician/Poet/Author (1809–1894)

64.30 Science moves, but slowly, slowly, creeping on from point to point.
Alfred, Lord Tennyson, English Poet (1809–1892)

64.31 Like truths of Science waiting to be caught.
Alfred, Lord Tennyson, English Poet (1809–1892)

64.32 Blessings on Science! When the earth seem'd old,
When Faith grew doting, and the Reason cold,
'Twas she discover'd that the world was young,
And taught a language to its lisping tongue:
'Twas she disclosed a future to its view,
And made old knowledge pale before the new.
Charles Mackay, Scottish Poet (1814–1889)

64.33 In our days everything seems pregnant with its contrary. Machinery, gifted with the wonderful power of shortening and fructifying human labor, we behold starving and overworking it. . . . At the same pace that mankind masters nature, man seems to become enslaved to other men or his own infamy. Even the pure light of science seems unable to shine but on the dark background of ignorance.
Karl Marx, German Political Philosopher (1818–1883)

64.34 Science deals exclusively with things as they are in themselves.
John Ruskin, English Art Critic/Author (1819–1900)

64.35 The work of science is to substitute facts for appearances, and demonstrations for impressions.
John Ruskin, English Art Critic/Author (1819–1900)

64.36 Science is, I verily believe, like virtue, its own exceeding great reward.

Charles Kingsley, English Poet/Novelist (1819–1875)

64.37 Science is for Life, not Life for Science.

Herbert Spencer, English Philosopher/Psychologist (1820–1903)

64.38 Science is organized knowledge.

Herbert Spencer, English Philosopher/Psychologist (1820–1903)

64.39 Only when genius is married to science, can the biggest results be produced.

Herbert Spencer, English Philosopher/Psychologist (1820–1903)

64.40 In science the important thing is to modify and change one's ideas as science advances.

Herbert Spencer, English Philosopher/Psychologist (1820–1903)

64.41 Science increases our power in proportion as it lowers our pride.

Herbert Spencer, English Philosopher/Psychologist (1820–1903)

64.42 Every science begins by accumulating observations, and presently generalizes these empirically; but only when it reaches the stage at which its empirical generalizations are included in a rational generalization does it become developed science.

Herbert Spencer, English Philosopher/Psychologist (1820–1903)

64.43 While natural science up to the end of the last century was predominantly a *collecting* science, a science of finished things, in our century it is essentially a *classifying* science, a science of processes, of the origin and development of these things and of the interconnection which binds these processes into one great whole.

Friedrich Engels, German Political Philosopher (1820–1895)

64.44 I consider the study of medicine to have been that training which preached more impressively and more convincingly than any other could have done, the everlasting principles of all scientific work; principles which are so simple and yet are ever forgotten again, so clear and yet always hidden by a deceptive veil.

Hermann Ludwig Ferdinand von Helmholtz, German Physicist/ Physiologist (1821–1894)

64.45 The task of science is to stake out the limits of the knowable, and to center consciousness within them.

Rudolph Virchow, German Pathologist/Statesman (1821–1902)

64.46 My kingdom is as wide as the world, and my desire has no limit. I go forward always, freeing spirits and weighing worlds, without fear, without compassion, without love, and without God. Men call me science.

Gustave Flaubert, French Novelist (1821–1880)

64.47 The Sciences gain by mutual support.
Louis Pasteur, French Chemist/Microbiologist (1822–1895)

64.48 Science is simply common sense at its best—that is, rigidly accurate in observation, and merciless to fallacy in logic.
Thomas Henry Huxley, English Biologist/Evolutionist (1825–1895)

64.49 Science is, I believe, nothing but trained and organized common sense, differing from the latter only as a veteran may differ from a raw recruit: and its methods differ from those of common sense only so far as the guardsman's cut and thrust differ from the manner in which a savage wields a club.
Thomas Henry Huxley, English Biologist/Evolutionist (1825–1895)

64.50 The generalizations of science sweep on in ever-widening circles, and more aspiring flights, through a limitless creation.
Thomas Henry Huxley, English Biologist/Evolutionist (1825–1895)

64.51 Science has fulfilled her function when she has ascertained and enunciated truth.
Thomas Henry Huxley, English Biologist/Evolutionist (1825–1895)

64.52 Every science has been an outcast.
Robert Green Ingersoll, American Lawyer (1833–1899)

64.53 Man has mounted science, and is now run away.
Henry (Brooks) Adams, American Historian/Author (1838–1918)

64.54 The sciences are beneficent. They prevent men from thinking.
Anatole France, French Novelist (1844–1924)

64.55 Science preceded the theory of science, and is independent of it. Science preceded naturalism, and will survive it.
Arthur James Balfour, English Statesman (1848–1930)

64.56 The work of science does not consist of creation but of the discovery of true thoughts.
Gottlob Frege, German Mathematician/Logician (1848–1925)

64.57 The extraordinary development of modern science may be her undoing. Specialism, now a necessity, has fragmented the specialties themselves in a way that makes the outlook hazardous. The workers lose all sense of proportion in a maze of minutiae.
Sir William Osler, Canadian Physician/Anatomist (1849–1919)

64.58 The future belongs to science. More and more she will control the destinies of the nations. Already she has them in her crucible and on her balances.
Sir William Osler, Canadian Physician/Anatomist (1849–1919)

64.59 Science, the new nobility! Progress. The world moves on!
Jean-Nicholas Arthur Rimbaud, French Poet (1854–1891)

64.60 Science is always wrong. It never solves a problem without creating ten more.
George Bernard Shaw, Irish Dramatist/Critic (1856–1950)

64.61 Every great advance in science has issued from a new audacity of the imagination.
John Dewey, American Psychologist/Educator/Philosopher (1859–1952)

64.62 Science is nothing but developed perception, interpreted intent, common sense rounded out and minutely articulated.
George Santayana, American Philosopher (1863–1952)

64.63 True science teaches, above all, to doubt and to be ignorant.
Miguel de Unamuno, Spanish Philosopher/Author (1864–1936)

64.64 Science? Pooh! Whatever good has science done the world? Damned bosh!
George Edward Moore, English Philosopher (1873–1958)

64.65 Science bestowed immense new powers on man, and, at the same time, created conditions which were largely beyond his comprehension.
Sir Winston Churchill, English Statesman/Author (1874–1965)

64.66 Equipped with his five senses, man explores the universe around him and calls the adventure Science.
Edwin Powell Hubble, American Astronomer (1889–1953)

64.67 Knowledge is a matter of science, and no dishonesty or conceit whatsoever is permissible.
Mao Tse Tung, Chinese Political Leader (1893–1976)

64.68 Moreover, in many of these rites we may discover the rudiments of science, the first gropings of man for an understanding of nature, and especially (as witchcraft is greatly concerned with the human body) the rudiments of medical science. In studying the very ignorance of primitive people with regard to nature, we are able to discern glimpses of real knowledge—we are, though not yet in the precincts, at any rate at the threshold of Science.
M. Winternitz, American Writer (1900)

64.69 Science is not a substitute for common sense, but an extension of it.
Willard van Orman Quine, American Philosopher (1908–)

64.70 A science is any discipline in which the fool of this generation can go beyond the point reached by the genius of the last generation.
Max Gluckman, American Anthropologist (1911–)

64.71 There are three great things in the world: there is religion, there is science, and there is gossip.

Robert Frost, American Poet (1963)

64.72 Out of man's mind in free play comes the creation Science. It renews itself, like the generations, thanks to an activity which is the best game of *homo ludens:* science is in the strictest and best sense a glorious entertainment.

Jacques Barzun, American Historian/Writer (1964)

64.73 Science 1. The state or fact of knowing; knowledge or cognizance *of* something specified or applied; also with wider reference, knowledge (more or less extensive) as a personal attribute.

***Oxford English Dictionary* (1971)**

64.74 Today it seems that people throw up their hands at the mention of science. It's become too complex, they say.

Christopher Hallowell, American Writer (1986)

65.
Science and Art

There is an art to science, and science

in art; the two are not enemies, but

different aspects of the whole.

ISAAC ASIMOV

65.1 For just as musical instruments are brought to perfection of clearness in the sound of their strings by means of bronze plates or horn sounding boards, so the ancients devised methods of increasing the power of the voice in theaters through the application of the science of harmony.

Marcus Vitruvius Pollio (Vitruvius), Roman Architect/Military Engineer (circa 100 B.C.)

65.2 The sciences and arts are not cast in a mold, but formed and shaped little by little, by repeated handling and polishing, as bears lick their cubs into shape at leisure.

Michel de Montaigne, French Essayist (1533–1592)

65.3 Science is the labor and handicraft of the mind; poetry can only be considered its recreation.

Francis Bacon, English Philosopher/Essayist/Statesman (1561–1626)

65.4 One science only will one genius fit,
So vast is art, so narrow human wit.

Alexander Pope, English Poet/Satirist (1688–1744)

65.5 There was far more imagination in the head of Archimedes than in that of Homer.

François-Marie Arouet de Voltaire, French Author/Philosopher (1694–1778)

65.6 The anatomist presents to the eye the most hideous and disagreeable objects, but his science is useful to the painter in delineating even a Venus or a Helen.

David Hume, Scottish Philosopher/Historian (1711–1776)

65.7 Science and art belong to the whole world, and the barriers of nationality vanish before them.

Johann Wolfgang von Goethe, German Poet/Dramatist/Novelist (1749–1832)

65.8 He who would do good to another must do it in Minute Particulars: General Good is the plea of the scoundrel, hypocrite, and flatterer, For Art and Science cannot exist but in minutely organized particulars.

William Blake, English Poet/Painter/Engraver (1757–1827)

65.9 Toil of science swells the wealth of art.

Friedrich von Schiller, German Poet/Dramatist (1759–1805)

65.10 I should not think of devoting less than twenty years to an epic poem. Ten years to collect materials and warm my mind with universal science. I would be a tolerable mathematician. I would thoroughly understand Mechanics; Hydrostatics; Optics and Astronomy; Botany; Metallurgy; Fossilism; Chemistry; Geology; Anatomy; Medicine; then the minds of men, in all Travels, Voyages, and Histories. So I would spend ten years; the next five in the composition of the poem, and the next five in the correction of it. So would I write, haply not unhearing of that divine and nightly whispering voice, which speaks to mighty minds, of predestined garlands, starry and unwithering.

Samuel Taylor Coleridge, English Poet/Critic (1772–1834)

65.11 Science has succeeded poetry, no less in the little walks of children than with men. Is there no possibility of averting this sore evil?

Charles Lamb, English Essayist/Critic (1775–1834)

65.12 Painting is a science and should be pursued as an inquiry into the laws of nature.

John Constable, English Painter (1776–1837)

65.13 In science, address the few, in literature, the many. In science, the few must dictate opinion to the many; in literature, the many, sooner or later, force their judgment on the few.

Edward Bulwer-Lytton, English Politician/Author (1803–1873)

65.14 What Art was to the ancient world, Science is to the modern.

Benjamin Disraeli, English Statesman/Writer (1804–1881)

65.15 Art is I; science is we.

Claude Bernard, French Physiologist (1813–1878)

65.16 We especially need imagination in science. It is not all mathematics, nor all logic, but it is somewhat beauty and poetry.

Maria Mitchell, American Astronomer (1818–1889)

65.17 This creative use of the imagination is not only the fountain of all inspiration in poetry and art, but is also the source of discovery in science, and indeed supplies the initial impulse to all development and progress. It is the creative power of imagination which has inspired and guided all the great discoveries in science.

Sir William Huggins, British Astronomer (1824–1910)

65.18 The man of science is nothing if not a poet gone wrong.

George Meredith, English Novelist/Poet (1828–1909)

65.19 Science is for those who learn; poetry, for those who know.

Joseph Roux, French Theologian/Epigrammist (1834–1886)

65.20 Mathematics possesses not only truth but supreme beauty—a beauty cold and austere, like that of a sculpture.

Bertrand Russell, English Philosopher/Mathematician (1872–1970)

65.21 Art upsets, science reassures.

Georges Braque, French Painter (1882–1963)

65.22 When it comes to atoms, language can be used only as in poetry. The poet, too, is not nearly so concerned with describing facts as with creating images.

Niels Henrik David Bohr, Danish Physicist (1885–1962)

65.23 Nobody, I suppose, could devote many years to the study of chemical kinetics without being deeply conscious of the fascination of time and change: this is something that goes outside science into poetry; but science, subject to the rigid necessity of always seeking closer approximations to the truth, itself contains many poetical elements.

Cyril Norman Hinshelwood, English Chemist (1897–1967)

65.24 The scientist has marched in and taken the place of the poet. But one day somebody will find the solution to the problems of the world and remember, it will be a poet, not a scientist.

Frank Lloyd Wright, American Architect (1959)

65.25 The two forms of experiment, scientific and artistic, share an attitude of detached yet intense observation toward the events precipitated. Yet the two activities are fundamentally opposed. *Scientific experiment*, relying on empirical rigor in refining its methods and verifying its results, seeks to extend and consolidate our grasp of order in the universe. *Artistic experiment* sets out to breed disorder, thwart determinism, and open up a space for individual freedom and consciousness.

Roger Shattuck, American Writer (1984)

65.26 Music critics have been known to disagree violently about the merits of a piece of music. Part of the reason may lie in the recent discovery that what one person hears can be strikingly different from what another hears. Some aspects of music appear to reside in the mind of the listener.

Ivars Peterson, American Science Journalist (1986)

66.
Science and Education

Science must be taught well, if a student is to understand the coming decades he must live through.

ISAAC ASIMOV

66.1 Much study is a weariness of the flesh.
***The Bible* (circa 725 B.C.)**

66.2 In addition to instructing them in the holy Scriptures, they also taught their pupils poetry, astronomy, and the calculation of the church calendar. [About the teaching methods of Theodore, Archbishop of Canterbury, and Hadrian, abbot of Canterbury, A.D. 669.]
Bede, English Monk/Scholar (673?–735)

66.3 For out of old feldes, as men seith,
Cometh al this newe corn fro yere to yere;
And out of olde bokes, in good feith,
Cometh al this newe science that men lere.
Geoffrey Chaucer, English Poet (1340?–1400)

66.4 If it were customary to send daughters to school like sons, and if they were then taught the natural sciences, they would learn as thoroughly and understand the subtleties of all the arts and sciences as well as sons. And by chance there happen to be such women, for, as I touched on before, just as women have more delicate bodies than men, weaker and less able to perform many tasks, so do they have minds that are freer and sharper whenever they apply themselves.

Christine de Pizan, French Writer/Poet (1365–1431)

66.5 Your father, who was a great scientist and philosopher, did not believe that women were worth less by knowing science; rather, as you know, he took great pleasure from seeing your inclination to learning. The feminine opinion of your mother, however, who wished to keep you busy with spinning and silly girlishness, following the common custom of women, was the major obstacle to your being more involved in the sciences. But your mother could not hinder in you the feeling for the sciences which you, through natural inclination, had nevertheless gathered together in little droplets.
[Describing her parents and herself.]

Christine de Pizan, French Writer/Poet (1365–1431)

66.6 I realize that Galen called an earth which contained metallic particles a mixed earth when actually it is a composite earth. But it behooves one who teaches others to give exact names to everything.

Georgius Agricola, German Physician/Politician/Mineralogist (1494–1555)

66.7 Histories make men wise; poets, witty; the mathematics, subtle; natural philosophy, deep, moral, grave; logic and rhetoric, able to contend.

Francis Bacon, English Philosopher/Essayist/Statesman (1561–1626)

66.8 And having thus passed the principles of arithmetic, geometry, astronomy, and geography, with a general compact of physics, they may descend in mathematics to the instrumental science of trigonometry, and from thence to fortification, architecture, engineering, or navigation. And in natural philosophy they may proceed leisurely from the history of meteors, minerals, plants, and living creatures, as far as anatomy. Then also in course might be read to them out of some not tedious writer the institution of physic. . . . To set forward all these proceedings in nature and mathematics, what hinders but that they may procure, as oft as shall be needful, the helpful experiences of hunters, fowlers, fishermen, shepherds, gardeners, apothecaries; and in other sciences, architects, engineers, mariners, anatomists.

John Milton, English Poet (1608–1674)

66.9 The business of the Society in their ordinary Meetings shall be, to order, take account, consider, and discourse of philosophical experiments and observations; to read, hear, and discourse upon letters, reports, and other papers, containing philosophical matters; as also to view, and discourse upon, rarities of nature and art; and thereupon to consider what may be deduced from them, or any of them; and how far they, or any of them, may be improved for use or discovery.

Statue of the Royal Society of London for the Promotion of Natural Knowledge (1663)

66.10 I had hitherto seen only one Side of the Academy, the other being appropriated to the Advancers of speculative Learning.

Some were condensing Air into a dry tangible Substance, by extracting the Nitre, and letting the acqueous or fluid Particles percolate. Others were softening Marble for Pillows and Pin-cushions. Another was, by a certain Composition of Gums, Minerals, and Vegetables outwardly applied, to prevent the Growth of Wool upon two young Lambs; and he hoped in a reasonable Time to propagate the Breed of naked Sheep all over the Kingdom.

Jonathan Swift, Irish Satirist/Clergyman (1667–1745)

66.11 The question [of the education of children] ought not to be to teach it the sciences, but to give it a taste for them, and methods to acquire them when the taste shall be better developed.

Jean-Jacques Rousseau, Swiss/French Philosopher/Author (1712–1778)

66.12 Science may be learned by rote, but Wisdom not.

Laurence Sterne, English Novelist (1713–1768)

66.13 But indeed, the English generally have been very stationary in latter times, and the French, on the contrary, so active and successful, particularly in preparing elementary books, in the mathematical and natural sciences, that those who wish for instruction, without caring from what nation they get it, resort universally to the latter language.

Thomas Jefferson, American President/Author (1743–1826)

66.14 Science has been seriously retarded by the study of what is not worth knowing, and of what is not knowable.

Johann Wolfgang von Goethe, German Poet/Dramatist/Novelist (1749–1832)

66.15 The lecturer should give the audience full reason to believe that all his powers have been exerted for their pleasure and instruction.

Michael Faraday, English Chemist/Physicist (1791–1867)

66.16 Every student who enters upon a scientific pursuit, especially if at a somewhat advanced period of life, will find not only that he has much to learn, but much also to unlearn.

Sir John Herschel, English Astronomer (1792–1871)

66.17 To discover and teach are functions; they are also distinct gifts, and are not commonly found united in the same person.

John Henry Newman, English Theologian/Author (1801–1890)

66.18 Science is the topography of ignorance.

Oliver Wendell Holmes, American Physician/Poet/Author (1809–1894)

66.19 In science, as in life, learning and knowledge are distinct, and the study of things, and not of books, is the source of the latter.

Thomas Henry Huxley, English Biologist/Evolutionist (1825–1895)

66.20 The child asks, "What is the moon, and why does it shine?" "What is this water and where does it run?" "What is this wind?" "What makes the waves of the sea?" "Where does this animal live, and what is the use of this plant?" And if not snubbed and stunted by being told not to ask foolish questions, there is no limit to the intellectual craving of a young child; nor any bounds to the slow, but solid, accretion of knowledge and development of the thinking faculty in this way. To all such questions, answers which are necessarily incomplete, though true as far as they go, may be given by any teacher whose ideas represent real knowledge and not mere book learning; and a panoramic view of Nature, accompanied by a strong infusion of the scientific habit of mind, may thus be placed within the reach of every child of nine or ten.

Thomas Henry Huxley, English Biologist/Evolutionist (1825–1895)

66.21 Truth is the aim of all research, no matter how sharply this truth may conflict with our social, ethical, and political conditions. This is the unifying bond of the modern university.

Christian Albert Theodor Billroth, Prussian Anatomist/Surgeon (1829–1894)

66.22 The present state of electrical science seems peculiarly unfavorable to speculation . . . to appreciate the requirements of the science, the student must make himself familiar with a considerable body of most intricate mathematics, the mere retention of which in the memory materially interferes with further progress. The first process therefore in the effectual study of the science, must be one of simplification and reduction of the results of previous investigation to a form in which the mind can grasp them.

James Clerk Maxwell, Scottish Physicist/Mathematician (1831–1879)

66.23 I'm very good at integral and differential calculus;
I know the scientific names of beings animalculous.

Sir W.S. (William Schwenk) Gilbert, English Comic Dramatist (1836–1911)

66.24 In gaining knowledge you must accustom yourself to the strictest sequence. You must be familiar with the very groundwork of science before you try to climb the heights. Never start on the "next" before you have mastered the "previous."

Ivan Petrovich Pavlov, Russian Physiologist (1849–1936)

66.25 I cannot answer your question, because I have not yet read that chapter in the textbook myself, but if you will come to me tomorrow I shall then have read it, and may be able to answer you.
[In reply to a student's question after a lecture.]

Jacques Loeb, German/American Zoologist/Biochemist (1879–1955)

66.26 We don't teach our students enough of the intellectual content of experiments—their novelty and their capacity for opening new fields. . . . My own view is that you take these things personally. You do an experiment because your own philosophy makes you want to know the result. It's too hard, and life is too short, to spend your time doing something because someone else has said it's important. You must feel the thing yourself.

Isidor Isaac Rabi, Austrian/American Physicist (1898–1963)

66.27 When you start in science, you are brainwashed into believing how careful you must be, and how difficult it is to discover things. There's something that might be called the "graduate student syndrome"; graduate students hardly believe they can make a discovery.

Francis Crick, British Biophysicist/Geneticist (1916–)

66.28 It is the task of general education to provide the kinds of learning and experience that will enable the student . . . to apply habits of scientific thought to both personal and civic problems, and to appreciate the implications of scientific discoveries for human welfare . . . [to] bring to the general student understanding of the fundamental nature of the physical world in which he lives and of the skills by which this nature is discerned.

The President's Commission on Higher Education (1947)

66.29 There is a story that once, not long after he came to Berlin, Planck forgot which room had been assigned to him for a lecture and stopped at the entrance office of the university to find out.

"Please tell me," he asked the elderly man in charge, "in which room does Professor Planck lecture today?"

The old man patted him on the shoulder. "Don't go there, young fellow," he said. "You are much too young to understand the lectures of our learned Professor Planck."

Barbara Lovett Cline, American Author (1965)

66.30 During the century after Newton, it was still possible for a man of unusual attainments to master all fields of scientific knowledge. But by 1800, this had become entirely impracticable.

Isaac Asimov, American Biochemist/Author (1965)

66.31 The scientific part of the curriculum [in 1250], the quadrivium, is not much influenced by the Greek science that scholars and translators are bringing in from the Moslem world. The pupil at the cathedral school absorbs relatively little true scientific knowledge. He may be given a smattering of natural history from the popular encyclopedias of the Dark Ages, based on Pliny and other Roman sources. He may learn, for example, that ostriches eat iron, that elephants fear only dragons and mice, that hyenas change their sex at will, that weasels conceive by the ear and deliver by the mouth.

Joseph and Frances Gies, American Historians (1973)

66.32 We really should be teaching brain science in high school.

Karl Pribram, Austrian/American Neurophysiologist (1984)

66.33 Maybe the situation is hopeless. Television is just the wrong medium, at least in prime time, to teach science. I think it is hopeless if it insists on behaving like television. . . . The people who produce these programs always respond to such complaints by insisting that no one would watch a program consisting of real scientists giving real lectures to real students. If they are right, then this sort of program is just another form of entertainment.

Jeremy Bernstein, American Physicist/Author (1986)

66.34 Since creationists hold that the universe is only ten thousand years old, this makes modern astronomy a joke. Since they reject radiometric dating as invalid, they contradict modern nuclear physics. Since they argue the Earth is only ten thousand years old as well, they are at odds with geology. Since they are offended by the human and animal family tree, they must oppose most of modern biochemistry.

Frederick Edwords, American Science Educator (1986)

66.35 Evidence abounds that we are graduating students who are unprepared and unable to grasp even day-to-day issues—the technical content of issues ranging from the safety of contraception devices or nutrition to robotics, gene-splicing, or the recent explosion of the space shuttle Challenger.

National Science Foundation, Conference on Science Education (1986)

66.36 One board member found perplexing the scientific view that the north and south poles have changed places several times in the history of our planet. She wondered why scientists couldn't make up their minds! She also found evolution unconvincing because she could not imagine how asexual life forms could ever evolve into male and female forms. Nonetheless, she said she was in favor of science because it had made possible cosmetics and skin creams.

National Science Foundation, Conference on Science Education (1986)

66.37 For the first time in our national history the higher-education enterprise that we pass on to our children and grandchildren will be less healthy, less able to respond to national needs . . . than the enterprise that we ourselves inherited.

David Allan Bromley, Canadian/American Physicist/Educator (1986)

66.38 The legislature has sent a clear message to us that they want the law on the books and they want us to defend it, and it's the Attorney General's obligation to carry that out.
[About a 1981 Louisiana state statute that requires the teaching of creationism as an alternative theory to evolution.]

Rusty Jabour, American Official for Louisiana Attorney General (1986)

66.39 The dispute between evolutionists and creation scientists offers textbook writers and teachers a wonderful opportunity to provide students with insights into the philosophy and methods of science. What students really need to know is how scientists judge the merit of a theory. Suppose students were taught the standards for evaluating a theory and then were asked to apply those criteria to the two theories in question. That would be an authentic education in science. . . .

We suspect that if the two theories were put side by side and students were given the freedom to judge their merit as science, creation theory would fall ignominiously. Natural selection is certainly not without flaws, but the point is that both bad science and bad education allow disputes over theory to go unexamined.

Neil Postman and Marc Postman, American Teacher and American Researcher (1986)

66.40 Today's children are being born into a technological world of computers. . . . if they can learn early on that they are in charge of that technology, then they are going to be more capable of dealing with the Information Age.

Victoria Williams and Frederick Williams, American Educator and Communications Professor (1986)

66.41 If more students don't get involved in science, this country may become a technological also-ran. . . . The once-numerous popular science magazines are folding, and school kids are turned off to science at an early age, just when they should be turned on to the unknown.

Christopher Hallowell, American Writer (1986)

66.42 That children no longer dream of flying to the stars has a lot of people worried, as it becomes quite clear that an increased reliance on science and technology *is* in the stars.

Christopher Hallowell, American Writer (1986)

67. Science and Journalism

Anyone who writes about science must know about science, which cuts down competition considerably.

ISAAC ASIMOV

67.1 Books must follow sciences, and not sciences books.
Francis Bacon, English Philosopher/Essayist/Statesman (1561–1626)

67.2 I have heard him [William Harvey] say that after his book of the Circulation of the Blood came out, that he fell mightily in his practice, and that 'twas believed by the vulgar that he was crack-brained.
John Aubrey, English Author/Biographer (1626–1697)

67.3 Johnson said that he could repeat a complete chapter of *The Natural History of Iceland* from the Danish of Horrebow, the whole of which was exactly thus: "There are no snakes to meet with throughout the whole island."
James Boswell, Scottish Lawyer/Author (1740–1795)

67.4 Work, Finish, Publish.
Michael Faraday, English Chemist/Physicist (1791–1867)

67.5 *Recollections* [his autobiographical work] might possibly interest my children or their children. I know that it would have interested me greatly to have read even so short and dull a sketch of the mind of my grandfather, written by himself, and what he thought and did, and how he worked. I have attempted to write the following account of myself, as if I were a dead man in another world looking back at my own life. Nor have I found this difficult, for life is nearly over with me.

Charles Robert Darwin, English Naturalist/Evolutionist (1809–1882)

67.6 When found, make a note of.

Charles Dickens, English Novelist (1812–1870)

67.7 When you believe you have found an important scientific fact, and are feverishly curious to publish it, constrain yourself for days, weeks, years sometimes, fight yourself, try and ruin your own experiments, and only proclaim your discovery after having exhausted all contrary hypotheses. But when, after so many efforts you have at last arrived at a certainty, your joy is one of the greatest which can be felt by a human soul.

Louis Pasteur, French Chemist/Microbiologist (1822–1895)

67.8 In science the credit goes to the man who convinces the world, not to the man to whom the idea first occurs.

Sir William Osler, Canadian Physician/Anatomist (1849–1919)

67.9 It seems better that a scientist should compile his own bibliography than that he should leave the work for another to do after he is dead.

Robert Broom, Scotch/South African Paleontologist (1866–1951)

67.10 Most of the fundamental ideas of science are essentially simple, and may, as a rule, be expressed in a language comprehensible to everyone.

Albert Einstein, German/American Physicist (1879–1955)

67.11 If you cannot—in the long run—tell everyone what you have been doing, your doing has been worthless.

Erwin Schrödinger, Austrian Physicist (1887–1961)

67.12 Our papers have been making a great deal of American "know-how" ever since we had the misfortune to discover the atomic bomb. There is one quality more important than "know-how" and we cannot accuse the United States of any undue amount of it. This is "know-what," by which we determine not only how to accomplish our purposes, but what our purposes are to be.

Norbert Wiener, American Inventor/Mathematician (1894–1964)

67.13 In the long run it pays the scientist to be honest, not only by not making false statements, but by giving full expression to facts that are opposed to his views. Moral slovenliness is visited with far severer penalties in the scientific than in the business world.

F. Cramer, American Scientist (1896)

67.14 I do not mind if you think slowly. But I do object when you publish more quickly than you think.

Wolfgang Pauli, German/American Physicist (1900–1958)

67.15 For some months the astronomer Halley and other friends of Newton had been discussing the problem in the following precise form: what is the path of a body attracted by a force directed toward a fixed point, the force varying in intensity as the inverse of the distance? Newton answered instantly, "An ellipse." "How do you know?" he was asked. "Why, I have calculated it." Thus originated the imperishable *Principia,* which Newton later wrote out for Halley. It contained a complete treatise on motion.

Eric Temple Bell, American Mathematician/Author (1937)

67.16 Even for the physicist the description in plain language will be a criterion of the degree of understanding that has been reached.

Werner Karl Heisenberg, German Physicist (1958)

67.17 When at a conference, he [Wolfgang Pauli] was introduced to Professor Paul Ehrenfest of Leiden University, a very distinguished physicist whose papers were greatly admired by his fellows and who himself was an admirer of Pauli's relativity article, Pauli was very rude. Ehrenfest told him quite frankly, "I like your publications better than I like you."

To which the young Pauli crushingly replied, "Strange. My feeling about you is just the opposite."

Barbara Lovett Cline, American Author (1965)

67.18 I think that we need to look at science journalism as a type of journalism, rather than a subfield of science.

Julie Ann Miller, American Science Journalist (1986)

68. Science and Mathematics

It is quite possible that mathematics was invented in the ancient Middle East to keep track of tax receipts and grain stores. How odd that out of this should come a subtle scientific language that can effectively describe and predict the most arcane aspects of the Universe.

ISAAC ASIMOV

68.1 Let no one ignorant of geometry enter my door.
Plato, Greek Philosopher (427 B.C.?–347 B.C.?)

68.2 For he who knows not mathematics cannot know any other science; what is more, he cannot discover his own ignorance, or find its proper remedy.
Roger Bacon, English Philosopher/Clergyman (1220–1292)

68.3 Mathematics is both the door and the key to the sciences.
Roger Bacon, English Philosopher/Clergyman (1220–1292)

68.4 The Universe is a grand book of philosophy. The book lies continually open to man's gaze, yet none can hope to comprehend it who has not first mastered the language and the characters in which it has been written. This language is mathematics; these characters are triangles, circles, and other geometric figures.

Galileo Galilei, Italian Astronomer/Physicist (1564–1642)

68.5 Geometry, which is the only science that it hath pleased God hitherto to bestow on mankind.

Thomas Hobbes, English Philosopher (1588–1679)

68.6 As I considered the matter carefully, it gradually came to light that all those matters only are referred to mathematics in which order and measurements are investigated, and that it makes no difference whether it be in numbers, figures, stars, sounds, or any other object that the question of measurement arises. I saw, consequently, that there must be some general science to explain that element as a whole which gives rise to problems about order and measurement. This I perceived was called universal mathematics. Such a science should contain the primary rudiments of human reason, and its province ought to extend to the eliciting of true results in every subject.

René Descartes, French Mathematician/Philosopher (1596–1650)

68.7 In mathematics we find the primitive source of rationality; and to mathematics must the biologists resort for means to carry out their researches.

Auguste Comte, French Philosopher (1798–1857)

68.8 We cannot take anything for granted, beyond the first mathematical formula. Question everything else.

Maria Mitchell, American Astronomer (1818–1889)

68.9 If you can measure that of which you speak, and can express it by a number, you know something of your subject; but if you cannot measure it, your knowledge is meager and unsatisfactory.

William Thomson, Lord Kelvin, British Physicist (1824–1907)

68.10 While the Mathematician is busy with deductions from general propositions, the Biologist is more especially occupied with observation, comparison, and those processes which lead *to* general propositions.

Thomas Henry Huxley, English Biologist/Evolutionist (1825–1895)

68.11 It may well be doubted whether, in all the range of science, there is any field so fascinating to the explorer—so rich in hidden treasures—so fruitful in delightful surprises—as that of Pure Mathematics. The charm lies chiefly . . . in the absolute certainty of its results; for that is what, beyond all mental treasures, the human intellect craves for. Let us only be sure of *something*! More light, more light!

Lewis Carroll (Charles Lutwidge Dodgson), English Author/ Mathematician (1832–1898)

68.12 "Can you do addition?" the White Queen said. "What's one and one and one and one and one and one and one and one and one and one?"
"I don't know," said Alice. "I lost count."
"She can't do addition," the Red Queen interrupted.
Lewis Carroll (Charles Lutwidge Dodgson), English Author/ Mathematician (1832–1898)

68.13 Oh these mathematicians make me tired! When you ask them to work out a sum they take a piece of paper, cover it with rows of A's, B's, and X's, Y's . . . scatter a mess of flyspecks over them, and then give you an answer that's all wrong.
Thomas Alva Edison, American Inventor (1847–1931)

68.14 The Science of Pure Mathematics . . . may claim to be the most original creation of the human spirit.
Alfred North Whitehead, English Mathematician/Philosopher (1861–1947)

68.15 Mathematics as a science commenced when first someone, probably a Greek, proved propositions about *any* things or about *some* things, without specification of definite particular things. These propositions were first enunciated by the Greeks for geometry; and, accordingly, geometry was the great Greek mathematical science.
Alfred North Whitehead, English Mathematician/Philosopher (1861–1947)

68.16 All science as it grows toward perfection becomes mathematical in its ideas.
Alfred North Whitehead, English Mathematician/Philosopher (1861–1947)

68.17 Mathematics may be defined as the subject in which we never know what we are talking about, nor whether what we are saying is true.
Bertrand Russell, English Philosopher/Mathematician (1872–1970)

68.18 Statistics are no substitute for judgment.
Henry Clay, American Politician (1879–1955)

68.19 There is a very significant characteristic of the application of the spiral to organic forms; that application invariably results in the discovery that nothing which is alive is ever simply mathematical. In other words, there is in every organic object a factor which baffles mathematics—a factor which we can only describe as Life.
Theodore Andrea Cook, English Engineer/Author (1914)

68.20 Statistics are like a bikini: what they reveal is suggestive but what they conceal is vital.
Anonymous (1986)

69.
Science and Philosophy

"Philosopher" is Greek for "lover of wisdom." How many students have longed for some philosophers to be, with equal dedication, "haters of obscurity."

ISAAC ASIMOV

69.1 The philosopher of science is not much interested in the thought processes which lead to scientific discoveries; he looks for a logical analysis of the completed theory, including the relationships establishing its validity. That is, he is not interested in the context of discovery, but in the context of justification.

Galileo Galilei, Italian Astronomer/Physicist (1564–1642)

69.2 Physical science will not console me for the ignorance of morality in the times of affliction. But the science of ethics will always console me for the ignorance of the physical sciences.

Blaise Pascal, French Mathematician/Philosopher/Author (1623–1662)

69.3 [Philosophy is] the science of sciences.

Samuel Taylor Coleridge, English Poet/Critic (1772–1834)

69.4 Science never saw a ghost, nor does it look for any, but it sees everywhere the traces, and it is itself the agent, of a Universal Intelligence.

Henry David Thoreau, American Writer/Naturalist (1817–1862)

69.5 The philosopher has no objections to a physicist's beliefs, so long as they are not advanced in the form of a philosophy.
Hans Reichenbach, German Philosopher (1861–1947)

69.6 I am now convinced that theoretical physics is actual philosophy.
Max Born, German/British Physicist (1882–1970)

69.7 We feel that even when *all possible* scientific questions have been answered, the problems of life remain completely untouched. Of course there are then no questions left, and this itself is the answer.
Ludwig Wittgenstein, Austrian Philosopher (1889–1951)

69.8 In saying that philosophy is concerned with each of the sciences . . . we mean also to rule out the supposition that philosophy can be ranged alongside the existing sciences, as a special department of speculative knowledge.
A.J. Ayer, British Philosopher (1910–)

69.9 It is evident that scientists and philosophers can help each other. For the scientist sometimes wants a new idea, and the philosopher is enlightened as to meanings by the study of the scientific consequences.
Alfred North Whitehead, English Mathematician/Philosopher (1938)

69.10 Metaphysics is a word which can mean exactly what one wants it to mean, hence its continuing popularity. To Aristotle it meant the field of speculation he took up after physics.
Roger Shattuck, American Writer (1984)

70.
Science and Politics

The nations may be divided in everything else, but they all share a single body of science.

ISAAC ASIMOV

70.1 And when statesman or others worry [the scientist] too much, then he should leave with his possessions. With a firm and steadfast mind one should hold under all conditions, that everywhere the earth is below and the sky above, and to the energetic man, every region is his fatherland.
Tycho Brahe, Danish Astronomer (1546–1601)

70.2 [We] do not learn for want of time,
The sciences that should become our country.
William Shakespeare, English Dramatist/Poet (1564–1616)

70.3 If science produces no better fruits than tyranny, murder, rapine, and destitution of national morality, I would rather wish our country to be ignorant, honest, and estimable as our neighboring savages are.
Thomas Jefferson, American President/Author (1743–1826)

70.4 Congress shall have power . . . to promote the progress of science and the useful arts by securing for limited times to . . . inventors the exclusive right to their . . . inventions.
[All patent acts were fostered by this declaration.]
United States Congress (1787)

70.5 About the year 1821, I undertook to superintend, for the Government, the construction of an engine for calculating and printing mathematical and astronomical tables. Early in the year 1833, a small portion of the machine was put together, and was found to perform its work with all the precision which had been anticipated. At that period circumstances, which I could not control, caused what I then considered a temporary suspension of its progress; and the Government, on whose decision the continuance or discontinuance of the work depended, have not yet communicated to me their wishes on the question.

Charles Babbage, English Inventor/Mathematician (1792–1871)

70.6 The Republic has no need of learned men.
[Reply to Antoine Laurent Lavoisier's request for two weeks' reprieve of his execution to finish his last experiment.]

Jean Baptiste Coffinhal, French Revolutionary (1794)

70.7 Politics is not an exact science.

Otto Eduard Leopold von Bismarck, German Statesman (1815–1898)

70.8 It is always observable that the physical and exact sciences are the last to suffer under despotisms.

Richard Henry Dana, Jr., American Lawyer/Author (1815–1882)

70.9 Science knows no country because knowledge belongs to humanity, and is the torch which illuminates the world. Science is the highest personification of the nation because that nation will remain the first which carries the furthest the works of thought and intelligence.

Louis Pasteur, French Chemist/Microbiologist (1822–1895)

70.10 The hostility of the state would be assumed toward any system or science that might not strengthen its arm.

Henry (Brooks) Adams, American Author/Historian (1838–1918)

70.11 Bourgeois scientists make sure that their theories are not dangerous to God or to capital.

Georgi Valentinovich Plekhanov, Russian Political Philosopher (1857–1918)

70.12 Under the flag of science, art, and persecuted freedom of thought, Russia would one day be ruled by toads and crocodiles the like of which were unknown even in Spain at the time of the Inquisition.

Anton Pavlovich Chekhov, Russian Dramatist/Author (1860–1904)

70.13 In holding scientific research and discovery in respect, as we should, we must be alert to the equal and opposite danger that public policy could itself become the captive of a scientific-technological elite.

Dwight David Eisenhower, American Soldier/President (1890–1969)

70.14 We can assuredly build a socialist state with modern industry, modern agriculture, and modern science and culture.

Mao Tse Tung, Chinese Political Leader (1893–1976)

70.15 Let both sides seek to invoke the wonders of science instead of its terrors. Together let us explore the stars, conquer the deserts, eradicate disease, tap the ocean depths, and encourage the arts and commerce.

John F. Kennedy, American President (1917–1963)

70.16 Science and Politics are forbidden Art's language of love, which would make us *trust* them; Politics and Art are forbidden Science's language of objectivity, which would make us *believe* them; and Art and Science are forbidden Politics' language of justice, which would move us to action. All these disjunctions tend to protect the *status quo* in our society.

Jay Lemke, American Anthropologist (1982)

70.17 Given the difficult decisions the President faces on both basic research and controversial military projects, three science advisors in six months does not inspire confidence.

American Association for the Advancement of Science (1986)

70.18 So it is no coincidence that the free nations have once again been the source of technological innovation. Once again an economic system congenial to free scientific inquiry, entrepreneurial risk-taking, and consumer freedom has been the fount of creativity and the mechanism for spreading innovation far and wide.

George Shultz, American Secretary of State (1986)

71.

Science and Religion

Religion cannot object to science on moral grounds. The history of religious intolerance forbids it.

ISAAC ASIMOV

71.1 He that increaseth knowledge, increaseth sorrow.
The Bible (circa 725 B.C.)

71.2 But considering all that has been said, it is possible to believe that the earth is so moved and that the heavens are not, and it is not evident to the contrary. And, in any case, it seems on the face of it, as much or more against natural reason, as are either all or some of the articles of our faith. And thus what I have said for fun in this manner could be of value in confuting or reproving those who would attack our faith by rational arguments.
Nicole Oresme, French Philosopher/Clergyman (1323–1382)

71.3 The humble knowledge of thyself is a surer way to God than the deepest search after science.
Thomas à Kempis, German Theologian/Clergyman (1380?–1471)

71.4 Thus, since it is not possible that the universe is enclosed between a material center and a circumference, the world is unintelligible; the universe whose center and circumference are God.
Nicholas of Cusa, German Cardinal/Mathematician/Statesman (1401–1464)

71.5 I, Galileo, son of the late Vincenzo Galilei, Florentine, aged seventy years, arraigned personally before this tribunal and kneeling before you, Most

Eminent and Reverend Lord Cardinal Inquisitor-General, against heretical pravity throughout the entire Christian commonwealth, having before my eyes and touching with my hands the Holy Gospels, swear that I have always believed, do believe, and by God's help will in the future believe all that is held, preached, and taught by the Holy Catholic and Apostolic Church. But, whereas—after an injunction had been judicially intimated to me by this Holy Office to the effect that I must altogether abandon the false opinion that the Sun is the center of the world and immovable and that the Earth is not the center of the world and moves and that I must not hold, defend, or teach in any way whatsoever, verbally or in writing, the said false doctrine . . .
[His Formula of Abjuration, June 22, 1633.]

Galileo Galilei, Italian Astronomer/Physicist (1564–1642)

71.6 I do not hope for any relief, and that is because I have committed no crime. I might hope for and obtain pardon, if I had erred, for it is to faults that the prince can bring indulgence, whereas against one wrongfully sentenced while he was innocent, it is expedient, in order to put up a show of strict lawfulness, to uphold rigor. . . . But my most holy intention, how clearly would it appear if some power would bring to light the slanders, frauds, and stratagems, and trickeries that were used eighteen years ago in Rome in order to deceive the authorities!
[From a letter to Pieresc, 1635.]

Galileo Galilei, Italian Astronomer/Physicist (1564–1642)

71.7 You have read my writings, and from them you have certainly understood which was the true and real motive that caused, under the lying mask of religion, this war against me that continually restrains and undercuts me in all directions, so that neither can help come to me from outside nor can I go forth to defend myself, there having been issued an express order to all Inquisitors that they should not allow any of my works to be reprinted which had been printed many years ago or grant permission to any new work that I would print. . . . a most rigorous and general order, I say, against all my works, *omnia et edenda;* so that it is left to me only to succumb in silence under the flood of attacks, exposures, derision, and insult coming from all sides.

Galileo Galilei, Italian Astronomer/Physicist (1564–1642)

71.8 So then Gravity may put the Planets into Motion, but without the divine Power it could never put them into such a circulating Motion as they have about the Sun; and therefore, for this, as well as other Reasons, I am compelled to ascribe the Frame of this System to an intelligent Agent.
[Letter from Newton to Richard Bently, January 17, 1693.]

Isaac Newton, English Physicist/Mathematician (1642–1727)

71.9 The Atoms or Particles, which now constitute Heaven and Earth, being once separate and diffused in the Mundane Space, like the supposed Chaos,

could never without a God by their Mechanical affections have convened into this present Frame of Things or any other like it.
[From "A Confutation of Atheism from the Origin and Frame of the World," 1693.]

Richard Bentley, English Clergyman/Author/Critic (1662–1742)

71.10 As to the Christian religion, Sir, we have a balance in its favor from the number of great men who have been convinced of its truth after a serious consideration of it. Grotius was an acute man, a lawyer, a man accustomed to examine evidence, and he was convinced; and he was no recluse man, but a man of the world, who surely had no bias towards it. Sir Isaac Newton set out as an infidel, but came to be a very firm believer.

Samuel Johnson, English Lexicographer/Poet/Critic (1709–1784)

71.11 Science is the great antidote to the poison of enthusiasm and superstition.

Adam Smith, Scottish Economist (1723–1790)

71.12 It is admitted, on all hands, that the Scriptures are not intended to resolve physical questions, or to explain matters in no way related to the morality of human actions; and if, in consequence of this principle, a considerable latitude of interpretation were not allowed, we should continue at this moment to believe, that the earth is flat; that the sun moves round the earth; and that the circumference of a circle is no more than three times its diameter.

John Playfair, Scottish Mathematician/Geologist (1748–1819)

71.13 The Atoms of Democritus
And Newton's Particles of Light
Are sands upon the Red Sea shore
Where Israel's tents do shine so bright.

William Blake, English Poet/Painter/Engraver (1757–1827)

71.14 To teach doubt and Experiment
Certainly was not what Christ meant.

William Blake, English Poet/Painter/Engraver (1757–1827)

71.15 Science surpasses the old miracles of mythology.

Ralph Waldo Emerson, American Essayist/Philosopher/Poet (1803–1882)

71.16 The narrow sectarian cannot read astronomy with impunity. The creeds of his church shrivel like dried leaves at the door of the observatory.

Ralph Waldo Emerson, American Essayist/Philosopher/Poet (1803–1882)

71.17 Science corrects the old creeds, sweeps away, with every new perception, our infantile catechisms, and necessitates a faith commensurate with the grander orbits and universal laws which it discloses.

Ralph Waldo Emerson, American Essayist/Philosopher/Poet (1803–1882)

71.18 In the endeavor to clearly comprehend and explain the functions of the combination of forces called "brain," the physiologist is hindered and troubled by the views of the nature of those cerebral forces which the needs of dogmatic theology have imposed on mankind.

Richard Owen, English Paleontologist/Biologist (1804–1892)

71.19 Science—in other words, knowledge—is not the enemy of religion; for, if so, then religion would mean ignorance. But it is often the antagonist of school-divinity.

Oliver Wendell Holmes, American Physician/Poet/Author (1809–1894)

71.20 I feel most deeply that the whole subject [of science and religion] is too profound for the human intellect. A dog might as well speculate on the mind of Newton. Let each man hope and believe what he can.

Charles Robert Darwin, English Naturalist/Evolutionist (1809–1882)

71.21 Geology, ethnology, what not?—
(Greek endings, each the little passing bell
That signifies some faith's about to die.)

Robert Browning, English Poet (1812–1889)

71.22 With all your science can you tell how it is, and whence it is, that light comes into the soul?

Henry David Thoreau, American Writer/Naturalist (1817–1862)

71.23 The superstition of science scoffs at the superstition of faith.

James Anthony Froude, English Historian (1818–1894)

71.24 Religion has been compelled by science to give up one after another of its dogmas, of those assumed cognitions which it could not substantiate. In the meantime, science substituted for the personalities to which religion ascribed phenomena certain metaphysical entities: and in doing this it trespassed on the province of religion; since it classed among the things which it comprehended certain forms of the incomprehensible.

Herbert Spencer, English Philosopher/Psychologist (1820–1903)

71.25 The task of science is . . . not to attach objects of faith, but to establish them and more convincingly than any other could have done, the everlasting principles of all scientific work; principles which are so simple and yet are ever forgotten again, so clear and yet always obscured by a deceptive veil.

Rudolf Virchow, German Pathologist/Statesman (1821–1902)

71.26 The birth of science was the death of superstition.

Thomas Henry Huxley, English Biologist/Evolutionist (1825–1895)

71.27 Extinguished theologians lie about the cradle of every science, as the strangled snakes besides that of Hercules.

Thomas Henry Huxley, English Biologist/Evolutionist (1825–1895)

71.28 I hardly know of a great physical truth whose universal reception has not been preceded by an epoch in which the most estimable persons have maintained that the phenomena investigated were directly dependent on the Divine Will, and that the attempt to investigate them was not only futile but blasphemous. And there is a wonderful tenacity of life about this sort of opposition to physical science. Crushed and maimed in every battle, it yet seems never to be slain; and after a hundred defeats it is at this day as rampant, though happily not so mischievous, as in the time of Galileo.

Thomas Henry Huxley, English Biologist/Evolutionist (1825–1895)

71.29 True science and true religion are twin sisters, and the separation of either from the other is sure to prove the death of both. Science prospers exactly in proportion as it is religious; and religion flourishes in exact proportion to the scientific depth and firmness of its bases.

Thomas Henry Huxley, English Biologist/Evolutionist (1825–1895)

71.30 Science seems to me to teach in the highest and strongest manner the great truth which is embodied in the Christian conception of entire surrender to the will of God. Sit down before fact as a little child, be prepared to give up every preconceived notion, follow humbly wherever and to whatever abysses nature leads, or you shall learn nothing. I have only begun to learn content and peace of mind since I have resolved at all risks to do this.

Thomas Henry Huxley, English Biologist/Evolutionist (1825–1895)

71.31 Science commits suicide when it adopts a creed.

Thomas Henry Huxley, English Biologist/Evolutionist (1825–1895)

71.32 The belief in the immortality of the human soul is a dogma which is in hopeless contradiction with the most solid empirical truths of modern science.

Ernst Heinrich Haeckel, German Biologist/Philosopher (1834–1919)

71.33 The next great task of Science is to create a religion for humanity.

John Morley, English Writer/Statesman (1838–1923)

71.34 If, finally, the science should prove that society at a certain time revert to the church and recover its old foundation of absolute faith in a personal providence and a revealed religion, it commits suicide.

Henry (Brooks) Adams, American Author/Historian (1838–1918)

71.35 By means of the Mummy, mankind, it is said,
Attests to the gods its respect for the dead.
We plunder his tomb, be he sinner or saint,
Distill him for physic and grind him for paint,
Exhibit for money his poor, shrunken frame,
And with levity flock to the scene of the shame.
O, tell me, ye gods, for the use of my rhyme:
For respecting the dead what's the limit of time?

Ambrose Bierce, American Satirist (1842–1914?)

71.36 Religion is comparable to a childhood neurosis.

Sigmund Freud, Austrian Psychiatrist/Psychoanalyst (1856–1939)

71.37 Those who speak of the incompatibility of science and religion either make science say that which it never said or make religion say that which it never taught.

Pope Pius XI (1857–1939)

71.38 The church saves sinners, but science seeks to stop their manufacture.

Elbert Hubbard, American Editor/Educator (1856–1915)

71.39 It is only through the psyche that we can establish that God acts upon us, but we are unable to distinguish whether these actions emanate from God or from the unconscious. We cannot tell whether God and the unconscious are two different entities. Both are border-line concepts for transcendental contents.

Carl Gustav Jung, Swiss Psychiatrist/Psychologist (1875–1961)

71.40 The scientific man often asserts that he cannot find God in Science.

F. Temple, English Philosopher (1885)

71.41 To study religions in a scientific spirit is to admit that all religions, if not equally good, spring at least from a common source.

***Encyclopaedia Brittanica* (1902)**

71.42 In our Jewish religious traditions today . . . we would never purport to place the Creation Epic as a scientific theory of creation. We understand it as a theological statement. . . . The truth of Adam and Eve stories, or of any other Biblical tales, does not rise or fall on their scientific demonstrability, but rather on their moral and symbolic teaching.

Rabbi Amiel Wohl, American Religious Leader (1876–1958)

71.43 Nor is there any evidence in the town of that poisonous spirit which usually shows itself when Christian men gather to defend the great doctrine of their faith. I have heard absolutely no whisper that Scopes is in the pay of the Jesuits, or that the whisky trust is backing him, or that he is egged on by the Jews who manufacture lascivious moving pictures. On the contrary, the Evolutionists and the Anti-Evolutionists seem to be on the best of terms and it is hard in a group to distinguish one from the other.
[About the Scopes Trial, Dayton, Ohio, 1925.]

H.L. Mencken, American Journalist/Critic (1880–1956)

71.44 Christ cannot possibly have been a Jew. I don't have to prove that scientifically. It is a fact.

Paul Joseph Goebbels, German Nazi Leader (1897–1945)

71.45 Dr. Oppenheimer grew tenser as the seconds ticked off . . . and then when the announcer shouted "now" and there came this tremendous burst of light followed immediately by the deep growling roar of the explosion, his face relaxed into an expression of tremendous relief. The effects could well be called unprecedented, magnificent, beautiful, stupendous, and terrifying. The explosion came . . . pressing hard against the people and things, to be followed almost immediately by the strong, sustained, awesome roar which warned of doomsday and made us feel that we puny things were blasphemous to dare tamper with the forces reserved for the Almighty.

War Department Release (1945)

71.46 I believe that the Dayton trial marked the beginning of the decline of fundamentalism. . . . I feel that restrictive legislation on academic freedom is forever a thing of the past, that religion and science may now address one another in an atmosphere of mutual respect and of a common quest for truth. I like to think that the Dayton trial had some part in bringing to birth this new era.

John T. Scopes, American High School Teacher (1965)

71.47 The Christian student of origins approaches the evidence from geology and paleontology with the Biblical record in mind, interpreting the evidence in accord with the facts of the Bible.

Kelly L. Segraves, American Creationist/Author (1975)

71.48 If entropy must constantly and continuously increase, then the universe is remorselessly running down, thus setting a limit (a long one, to be sure) on the existence of humanity. To some human beings, this ultimate end poses itself almost as a threat to their personal immortality, or as a denial of the omnipotence of God. There is, therefore, a strong emotional urge to deny that entropy *must* increase.

Isaac Asimov, American Biochemist/Author (1976)

71.49 I could almost wish, at this point, that I were in the habit of expressing myself in theological terms, for if I were, I might be able to compress my entire thesis into a sentence.

All knowledge of every variety (I might say) is in the mind of God—and the human intellect, even the best, in trying to pluck it forth can but "see through a glass, darkly."

Isaac Asimov, American Biochemist/Author (1976)

71.50 Why do we want to see creation-science in public schools? First, we feel that students have the right to know. At present, few students are exposed to the weaknesses of evolution, let alone to the data supporting the creation-science alternative. Including creation-science in a balanced approach would keep both positions honest.

George E. Hahn, American Creationist (1982)

71.51 If during the course of a person's education he is introduced to the principles of a religious attitude to life, then it is perhaps possible to learn to equate human rights with human obligation. It is possible then to see a relationship between moral values and science.

Charles, Prince of Wales, British Royalty (1986)

71.52 Liberals, the "good guys" in this matter, sound much like those "bad guys," who in 1925 prohibited by law the teaching of evolution in Tennessee. In the earlier case, anti-evolutionists feared that a scientific idea would undermine religious belief. In the present case, pro-evolutionists fear that a religious idea will undermine scientific belief. The former had insufficient confidence in religion; the latter, insufficient confidence in science.

Neil Postman and Marc Postman, American Teacher and American Researcher (1986)

72.
Science and Society

The saddest aspect of life right now is

that science gathers knowledge faster than

society gathers wisdom.

ISAAC ASIMOV

72.1 Nothing has tended more to retard the advancement of science than the disposition in vulgar minds to vilify what they cannot comprehend.
Samuel Johnson, English Lexicographer/Poet/Critic (1709–1784)

72.2 There prevails among men of letters an opinion, that all appearance of science is particularly hateful to women.
Samuel Johnson, English Lexicographer/Poet/Critic (1709–1784)

72.3 Society is, indeed, a contract. . . . It is a partnership in all science; a partnership in all art, a partnership in every virtue, and in all perfection. As the ends of such a partnership cannot be obtained in many generations, it becomes a partnership not only between those who are living, but between those who are living, those who are dead, and those who are to be born.
Edmund Burke, English Statesman/Author (1729–1797)

72.4 The main object of all science is the freedom and happiness of man.
Thomas Jefferson, American President/Author (1743–1826)

72.5 'Tis a pity learned virgins ever wed
With persons of no sort of education,
Or gentlemen, who, though well born and bred,
Grow tired of scientific conversation.
Lord Byron (George Gordon), English Poet/Dramatist (1788–1824)

72.6 It is a matter of primary importance in the cultivation of those sciences in which truth is discoverable by the human intellect that the investigator should be free, independent, unshackled in his movement; that he should be allowed and enabled to fix his mind intently, nay, exclusively, on his special object, without the risk of being distracted every other minute in the process and progress of his inquiry by charges of temerariousness, or by warnings against extravagance or scandal.

John Henry Newman, English Theologian/Author (1801–1890)

72.7 Where speculation ends—in real life—there real, positive science begins: the representation of the practical activity, of the practical process of development of men. Empty talk about consciousness ceases, and real knowledge has to take its place.

Karl Marx, German Political Philosopher (1818–1883)

72.8 The love of science, and the energy and honesty in the pursuit of science, in the best of the Aryan races, seem to correspond in a remarkable way to the love of conduct, and the energy and honesty in the pursuit of conduct, in the best of the Semitic.

Matthew Arnold, English Poet/Critic (1822–1888)

72.9 Some day science may have the existence of man in its power, and the human race may commit suicide by blowing up the world.

Henry (Brooks) Adams, American Author/Historian (1838–1918)

72.10 Who can doubt that the leaven of science, working in the individual, leavens in some slight degree the whole social fabric? Reason is at least free, or nearly so; the shackles of dogma have been removed, and faith herself, freed from a morganatic alliance, finds in the release great gain.

Sir William Osler, Canadian Physician/Anatomist (1849–1919)

72.11 And so the great truth, now a paradox, may become a commonplace, that man is greater than his surroundings, and that the production of a breed of men and women, even in our great cities, less prone to disease, and pain, more noble in aspect, more rational in habits, more exultant in the pure joy of living, is not only scientifically possible, but that even the partial fulfillment of this dream, if dream it be, is the most worthy object towards which the lover of his kind can devote the best energies of his life.

Hely Hutchinson Almond, British Writer (1900)

72.12 Humanity has in the course of time had to endure from the hands of science two great outrages upon its self-love. The first was when it realized that our earth was not the center of the universe. . . . The second was when geological research robbed man of his peculiar privilege of having been specially created, and relegated him to descent from the animal world, implying an irradicable animal nature in him.

Sigmund Freud, Austrian Psychiatrist/Psychoanalyst (1856–1939)

72.13 Why does this magnificent applied science which saves work and makes life easier bring us so little happiness? The simple answer runs: because we have not yet learned to make sensible use of it.
Albert Einstein, German/American Physicist (1879–1955)

72.14 "Science says," will generally be found to settle any argument in a social gathering or sell any article from tooth-paste to refrigerators.
***The Nation* (1928)**

72.15 The persistence of race prejudice where it exists is a cultural acquisition, which as we have seen finds no justification in biology.
L.C. Dunn, American Geneticist (1951)

72.16 Science may be a boon if war can be abolished and democracy and cultural liberty preserved. If this cannot be done, science will precipitate evils greater than any that mankind has ever experienced.
Bertrand Russell, English Philosopher/Mathematician (1952)

72.17 The results of science, in the form of mechanism, poison gas, and the yellow press, bid fair to lead to the total downfall of our civilization.
Bertrand Russell, English Philosopher/Mathematician (1952)

72.18 What the scientists have in their briefcases is terrifying.
Nikita Sergeevich Khrushchev, Soviet Statesman/Prime Minister (circa 1960)

72.19 The language of science is universal, and perhaps scientists have been the most international of all professions in their outlook. But the contemporary revolution in transport and communication has dramatically speeded the internationalization of science.
John F. Kennedy, American President (circa 1960)

72.20 Science contributes to our culture in many ways, as a creative intellectual activity in its own right, as the light which has served to illuminate man's place in the universe, and as the source of understanding of man's own nature.
John F. Kennedy, American President (circa 1960)

72.21 You have only to wish it and you can have a world without hunger, disease, cancer, and toil—anything you wish, wish anything and it can be done.
Albert Szent-Gyorgyi, American Biochemist (1970)

72.22 Science promised man power. . . . But, as so often happens when people are seduced by promises of power, the price is servitude and impotence. Power is nothing if it is not the power to choose.
Dr. Joseph Weizenbaum, American Computer Scientist/Sociologist (1976)

72.23 If we are to avoid witch-hunting we will have to put drug testing back into the hands of medical diagnosticians, where it belongs. . . . Urine and blood do not speak any more scientifically than witches' marks.
Ronald K. Siegel, American Psychopharmacologist (1986)

73.
Scientific Apparatus

Scientific apparatus offers a window to knowledge, but as they grow more elaborate, scientists spend ever more time washing the windows.

ISAAC ASIMOV

73.1 What, then, shall we say about the receipts of alchemy, and about the diversity of its vessels and instruments? These are furnaces, glasses, jars, waters, oils, limes, sulphurs, salts, saltpeters, alums, vitriols, chrysocollae, copper greens, atraments, auripigments, fel vitri, ceruse, red earth, thucia, wax, lutum sapientiae, pounded glass, verdigris, soot, crocus of Mars, soap, crystal, arsenic, antimony, minium, elixir, lazarium, gold leaf, salt niter, sal ammoniac, calamine stone, magnesia, bolus armenus, and many other things. Then, again, concerning herbs, roots, seeds, woods, stones, animals, worms, bone dust, snail shells, other shells, and pitch. These and the like, whereof there are some very farfetched in alchemy, are mere incumbrances of work; since even if Sol and Luna [gold and silver] could be made by them they rather hinder and delay than further one's purpose.

Paracelsus (Theophrastus Bombastius) Swiss/German Chemist/ Physician (1493–1541)

73.2 Now I should like to ask you for an observation; since I possess no instruments, I must appeal to others.

Johannes Kepler, German Astronomer/Mathematician (1571–1630)

73.3 I can certainly wish for new, large, and properly constructed instruments, and enough of them, but to state where and by what means they are to be procured, this I cannot do. Tycho Brahe has given Mastlin an instrument of metal as a present, which would be very useful if Mastlin could afford the cost of transporting it from the Baltic, and if he could hope that it would travel such a long way undamaged. . . . One can really ask for nothing better for the observation of the sun than an opening in a tower and a protected place underneath.

Johannes Kepler, German Astronomer/Mathematician (1571–1630)

73.4 This Excellent Mathematician having given us, in the Transactions of February last, an account of the cause, which induced him to think upon Reflecting Telescopes, instead of Refracting ones, hath thereupon presented the curious world with an Essay of what may be performed by such Telescopes; by which it is found, that Telescopical Tubes may be considerably shortened without prejudice to their magnifying effect.
[About his invention of the catadioptrical telescope, in a communication to the *Philosophical Transactions,* March 25, 1672.]

Isaac Newton, English Physicist/Mathematician (1642–1727)

73.5 When Aloisio Galvani first stimulated the nervous fiber by the accidental contact of two heterogeneous metals, his contemporaries could never have anticipated that the action of the voltaic pile would discover to us, in the alkalies, metals of a silvery luster, so light as to swim on water, and eminently inflammable; or that it would become a powerful instrument of chemical analysis, and at the same time a thermoscope and a magnet.

Alexander von Humboldt, German Naturalist/Statesman (1769–1859)

73.6 May every young scientist remember . . . and not fail to keep his eyes open for the possibility that an irritating failure of his apparatus to give consistent results may once or twice in a lifetime conceal an important discovery.

Patrick Maynard Stuart Blackett, British Physicist (1897–1974)

73.7 It is important to realize that the basic ideas for instrumentation come from research in pure science. Our technology depends on instrumentation. The demands of our technology cause instruments to be available which, although they were designed for industrial use, often make research in pure science much easier. Thus, there is a fundamental relationship among pure science, instrumentation, and technology, with each benefiting from the others.

Ralph H. Munch, American Chemist (1952)

73.8 To make still bigger telescopes will be useless, for the light absorption and temperature variations of the earth's atmosphere are what now limits the ability to see fine detail. If bigger telescopes are to be built, it will have to be for use in an airless observatory, perhaps an observatory on the moon.

Isaac Asimov, American Biochemist/Author (1965)

73.9 We look forward to it as being the greatest scientific instrument that man ever built.
[On the 2.4-meter Hubble space telescope.]

Samuel W. Keller, American Administrator for NASA (1986)

73.10 In spite of a lot of stuff printed about how the Space Age is here and astronomers will never observe from the ground again, it really isn't all that simple. As long as you've got to shoot into space something the height of a thirty-six story building, it's going to be very expensive. That means there will still be a lot of instrument development and observations done here on Earth.

Robert Smithson, American Astronomer (1986)

74.
Scientific Funding

Every hour a scientist spends trying to raise funds is an hour lost from important thought and research.

ISAAC ASIMOV

74.1 August 29, 1662. The council and fellows of the Royal Society went in a body to Whitehall to acknowledge his Majesty's royal grace to granting our charter and vouchsafing to be himself our founder; then the president gave an eloquent speech, to which his Majesty gave a gracious reply and we all kissed his hand. Next day, we went in like manner with our address to my Lord Chancellor, who had much prompted our patent.

John Evelyn, English Diarist (1620–1706)

74.2 The loss of my pension [from the Academy of Sciences] . . . and the enormous increase in the price of articles of subsistence have placed me and my numerous family in a state of distress that leaves me neither the time nor the freedom from care to cultivate science in a useful way.

Jean-Baptiste de Monet, Chevalier de Lamarck, French Naturalist (1744–1829)

74.3 Between twenty and thirty of the most erudite citizens decided upon forming a phrenological society. A meeting was called, and fully attended; a respectable number of subscribers' names was registered, the payment of subscriptions being arranged for a future day. President, vice-president, treasurer, and secretary were chosen, and the first meeting dissolved with every appearance of energetic perseverance in scientific research. The second meeting brought together one half of this learned body, and they enacted rules and laws, and passed resolutions, sufficient, it was said, to have filled three folios. A third day of meeting arrived, which was an important one, as on this occasion the subscriptions were to be paid. The treasurer came punctually, but found himself alone. With patient hope, he waited for two hours for the wise men of the West, but he waited in vain: and so expired the Phrenological Society of Cincinnati.

Frances Trollope, English Writer (1780–1863)

74.4 Two years ago I tried to appeal to Rockefeller's conscience about the absurd method of allocating grants, unfortunately without success. Bohr has now gone to see him, in an attempt to persuade him to take some action on behalf of the exiled German scientists.

Albert Einstein, German/American Physicist (1879–1955)

74.5 If you find nothing, you are never to come begging [at my] door again. [To his son, paleontologist Richard Leakey, regarding Louis and Mary Leakey's funding of Richard's first dig in Kenya.]

Louis Leakey, British Paleontologist (1903–1972)

74.6 I had underestimated the business ability of my brown collectors. Behind my back they broke the larger fragments into pieces to increase the number of sales.
[Regarding his unsatisfactory arrangement with native workers in Sangiran to pay ten cents for each fossil fragment found.]

G.H.R. von Koenigswald, German Paleontologist (1953)

74.7 In the late 1950s a small number of scientists were conducting research with a recently invented device called the *maser* which was used to produce microwaves. The letters that made up the word stood for "Microwave Amplification by Stimulated Emission of Radiation." At least that was the official version; according to a joke current at the time, they really stood for "Money Acquisition Scheme for Expensive Research."

Richard Morris, American Physicist/Author (1979)

74.8 Eighty-five percent of the American scientists who have been stuck at the level of research associate for twenty-five years or more are women. Research associates are Ph.D.s with a lot of responsibility who are never really boss, calling the tune. On paper, you work for this excellent person who received grant money, but you're not good enough to be on the faculty yourself.

Candace Pert, American Neuroscientist (1984)

74.9 Economists use the expression "opportunity costs" for losses incurred through certain choices made over others, including ignorance and inaction. For systematics, or more precisely the neglect of systematics and the biological research dependent upon it, the costs are very high.

Edward O. Wilson, American Entomologist/Sociobiologist (1985)

74.10 The worldwide support for basic tropical biology, including systematics and ecology, is only about $50 million. Just 1,000 annual grants of $50,000 devoted to tropical organisms would double the level of support and revitalize the field. . . . The same amount added to the approximately $3.5 billion spent on health-related biology in the United States would constitute an increment of 1.4%, causing a barely detectable change.

Edward O. Wilson, American Entomologist/Sociobiologist (1985)

74.11 That bill, now law, directs the Department of Defense [DOD] to spend part of its current research budget to fund projects at ten specific universities. Although the instructions in the House report accompanying the spending bill refer to research, most of the funds, seemingly randomly distributed among several DOD offices, are to be used for equipment purchases and building construction.

[On a spending bill passed by Congress at the end of the 1985 session.]

Ivars Peterson, American Science Journalist (1986)

75. Scientific Method

It is hard to describe the exact route to scientific achievement, but a good scientist doesn't get lost as he travels it.

ISAAC ASIMOV

75.1 Plainly, then, these are the causes, and this is how many they are. They are four, and the student of nature should know them all, and it will be his method, when stating on account of what, to get back to them all: the matter, the form, the thing which effects the change, and what the thing is for.

Aristotle, Greek Philosopher (384 B.C.–322 B.C.)

75.2 In all disciplines in which there is systematic knowledge of things with principles, causes, or elements, it arises from a grasp of those: we think we have knowledge of a thing when we have found its primary causes and principles, and followed it back to its elements. Clearly, then, systematic knowledge of nature must start with an attempt to settle questions about principles.

Aristotle, Greek Philosopher (384 B.C.–322 B.C.)

75.3 The method of definition is the method of discovering what the thing under consideration is by means of the definition of that thing in so far as it makes it known. This method involves two procedures, one being by composition and the other by resolution.

Robert Grosseteste, English Bishop/Educator (1168–1253)

75.4 They are ill discoverers that think there is no land, when they can see nothing but sea.

Francis Bacon, English Philosopher/Essayist/Statesman (1561–1626)

75.5 For it is too bad that there are so few who seek the truth and so few who do not follow a mistaken method in philosophy. This is not, however, the place to lament the misery of our century, but to rejoice with you over such beautiful ideas for proving the truth. So I add only, and I promise, that I shall read your book at leisure; for I am certain that I shall find the noblest things in it. And this I shall do the more gladly, because I accepted the view of Copernicus many years ago, and from this standpoint I have discovered from their origins many natural phenomena, which doubtless cannot be explained on the basis of the more commonly accepted hypothesis.
[To Johannes Kepler.]

Galileo Galilei, Italian Astronomer/Physicist (1564–1642)

75.6 The discoveries that one can make with the microscope amount to very little, for one sees with the mind's eye and without the microscope the real existence of all these little beings.

Georges-Louis Leclerc, Comte de Buffon, French Naturalist (1707–1788)

75.7 As systematic unity is what first raises ordinary knowledge to the rank of science, that is, makes a system out of a mere aggregate of knowledge, architectonic [the art of constructing systems] is the doctrine of the scientific in our knowledge, and therefore necessarily forms part of the doctrine of method.

Immanuel Kant, German Philosopher (1724–1804)

75.8 My Lord said that he who knew men only in this way [from history] was like one who had got the theory of anatomy perfectly, but who in practice would find himself very awkward and liable to mistakes. That he again who knew men by observation was like one who picked up anatomy by practice, but who like all empirics would for a long time be liable to gross errors.

James Boswell, Scottish Lawyer/Author (1740–1795)

75.9 I am pleased, however, to see the efforts of hypothetical speculation, because by the collisions of different hypotheses, truth may be elicited and science advanced in the end.

Thomas Jefferson, American President/Author (1743–1826)

75.10 A patient pursuit of facts, and cautious combination and comparison of them, is the drudgery to which man is subjected by his Maker, if he wishes to attain sure knowledge.

Thomas Jefferson, American President/Author (1743–1826)

75.11 We must trust to nothing but facts: these are presented to us by Nature, and cannot deceive. We ought, in every instance, to submit our reasoning to the test of experiment, and never to search for truth but by the natural road of experiment and observation.

Antoine Lavoisier, French Chemist (1743–1794)

75.12 Man is naturally metaphysical and arrogant, and is thus capable of believing that the ideal creations of his mind, which express his feelings, are identical with reality. From this it follows that the experimental method is not really natural to him.

Claude Bernard, French Physiologist (1813–1878)

75.13 The experimenter who does not know what he is looking for will never understand what he finds.

Claude Bernard, French Physiologist (1813–1878)

75.14 Good methods can teach us to develop and use to better purpose the faculties with which nature has endowed us, while poor methods may prevent us from turning them to good account. Thus the genius of inventiveness, so precious in the sciences, may be diminished or even smothered by a poor method, while a good method may increase and develop it.

Claude Bernard, French Physiologist (1813–1878)

75.15 With acurate experiment and observation to work upon, imagination becomes the architect of physical theory.

John Tyndall, English Physicist (1820–1893)

75.16 Accurate and minute measurement seems to the nonscientific imagination a less lofty and dignified work than looking for something new. But nearly all the grandest discoveries of science have been but the rewards of accurate measurement—patient long-continued labor in the minute sifting of numerical results.

William Thomson, Lord Kelvin, British Physicist(1824–1907)

75.17 The method of scientific investigation is nothing but the expression of the necessary mode of working of the human mind. It is simply the mode at which all phenomena are reasoned about, rendered precise and exact.

Thomas Henry Huxley, English Biologist/Evolutionist (1825–1895)

75.18 It was a great step in science when men became convinced that, in order to understand the nature of things, they must begin by asking, not whether a thing is good or bad, noxious or beneficial, but of what kind it is? And how much is there of it? Quality and quantity were then first recognized as the primary features to be discovered in scientific inquiry.

James Clerk Maxwell, Scottish Physicist/Mathematician (1831–1879)

75.19 There is one thing even more vital to science than intelligent methods; and that is, the sincere desire to find out the truth, whatever it may be.

Charles Sanders Peirce, American Philosopher/Logician (1839–1914)

75.20 The really valuable factor is intuition.

Albert Einstein, German/American Physicist (1879–1955)

75.21 Every method is imperfect.
Charles-Jean-Henri Nicolle, French Physician/Bacteriologist (1932)

75.22 That the student grasp the processes involved in scientific thought and understand the principles of scientific method is even more important than that he should know the dates of the sciences.
The President's Commission on Higher Education (1947)

75.23 What we observe is not nature itself, but nature exposed to our method of questioning.
Werner Karl Heisenberg, German Physicist (1958)

75.24 The real purpose of scientific method is to make sure Nature hasn't misled you into thinking you know something you don't actually know. There's not a mechanic or scientist or technician alive who hasn't suffered from that one so much that he's not instinctively on guard. . . . If you get careless or go romanticizing scientific information, giving it a flourish here and there, Nature will soon make a complete fool out of you.
Robert Pirsig, American Author (1974)

75.25 Typically, scientific discovery is a two-part process. The first thing that happens is that a scientist experiences a sudden insight. Then, if he is lucky he finds that the insight has logical consequences that will clear up an outstanding scientific problem, or explain baffling experimental results.
Richard Morris, American Physicist/Author (1983)

75.26 I've noticed that most good research (and I hope mine is good) reflects the researcher's private questions. It took me a long time to see that my work grew out of my own concerns.
June Reinisch, American Director of the Kinsey Institute (1986)

75.27 The cause of the problem is often prior solutions.
Amory Lovins, American Physicist (1986)

75.28 The scientist, in practicing the scientific method, cannot utter a single word about an individual thing or creature insofar as it is an individual, but only insofar as it resembles individuals. This limitation holds true whether the individual is a molecule of NaCl or an amoeba or a human being. . . . We have no particular interest in this particular pinch of salt or this particular dogfish.
Walker Percy, American Novelist (1986)

76.
Scientists

A scientist is as weak and human as any man, but the pursuit of science may ennoble him even against his will.

ISAAC ASIMOV

76.1 So the astronomer is on common ground with the physicist both in the subject and in the predicate of the conclusion, but the physicist demonstrates the predicate to belong to the subject by nature, whereas the astronomer does not care whether it belongs by nature or not. What, therefore, is the predicate for the physicist, is abstracted as the subject for the pure mathematician.
Robert Grosseteste, English Bishop/Educator (1168–1253)

76.2 What others strive to see dimly and blindly, like bats in twilight, he [Petrus Peregrinus] gazes at in the full light of day, because he is a master of experiment. Through experiment he gains knowledge of natural things, medical, chemical, and indeed of everything in the heavens or earth. . . . He has even taken note of the remedies, lot casting, and charms used by old women and by wizards and magicians, and of the deceptions and devices of conjurors, so that nothing which deserves inquiry should escape him, and that he may be able to expose the falsehoods of magicians.
Roger Bacon, English Philosopher/Clergyman (1220–1292)

76.3 Let me not seem to have lived in vain.
[His last words.]
Tycho Brahe, Danish Astronomer (1546–1601)

76.4 I have written many direct and indirect arguments for the Copernican view, but until now I have not dared to publish them, alarmed by the fate of

Copernicus himself, our master. He has won for himself undying fame in the eyes of a few, but he has been mocked and hooted at by an infinite multitude (for so large is the number of fools). I would dare to come forward publicly with my ideas if there were more people of your [Johannes Kepler's] way of thinking. As this is not the case, I shall refrain.

Galileo Galilei, Italian Astronomer/Physicist (1564–1642)

76.5 Tycho [Brahe] is a man with whom no one can live without exposing himself to the greatest indignities. The pay is splendid, but one can only extract the half of it. I have thought of turning to medicine.

Johannes Kepler, German Astronomer/Mathematician (1571–1630)

76.6 I much prefer the sharpest criticism of a single intelligent man to the thoughtless approval of the great masses.

Johannes Kepler, German Astronomer/Mathematician (1571–1630)

76.7 Science distinguishes a Man of Honor from one of those Athletic Brutes whom undeservedly we call Heroes.

John Dryden, English Poet/Critic/Dramatist (1631–1700)

76.8 If I have seen a little further it is by standing on the shoulders of Giants.

Isaac Newton, English Physicist/Mathematician (1642–1727)

76.9 To myself I seem to have been only like a boy playing on the seashore, and diverting myself in now and then finding a smoother pebble or a prettier shell than ordinary, whilst the great ocean of truth lay all undiscovered before me.

Isaac Newton, English Physicist/Mathematician (1642–1727)

76.10 This incomparable Author having at length been prevailed upon to appear in public, has in this Treatise given a most notable instance of the extent of the powers of the Mind; and has at once shown what are the Principles of Natural Philosophy, and so far derived from them their consequences, that he seems to have exhausted his Argument, and left little to be done by those that shall succeed him.

[From a review of Newton's "Principia," in the *Philosophical Transactions*, 1687.]

Edmund Halley, English Astronomer (1656–1742)

76.11 Here lies Sir Isaac Newton, Knight, who by a vigour of mind almost supernatural, first demonstrated, the motions and Figures of the Planets, the Paths of the comets, and the Tides of the Oceans. . . . Let Mortals rejoice that there has existed such and so great an ornament of Nature.

Epitaph of Isaac Newton, on his tomb at Westminster Abbey (1727)

76.12 Go wondrous creature! mount where Science guides;
Go, measure earth, weigh air, and state the tides;

Instruct the planets in what orbs to run,
Correct old Time, and regulate the sun . . .
Go, teach Eternal Wisdom how to rule—
Then drop into thyself and be a fool!
Alexander Pope, English Poet/Satirist (1688–1744)

76.13 Nature and Nature's laws lay hid in night.
God said, "Let Newton be!" and all was light.
Alexander Pope, English Poet/Satirist (1688–1744)

76.14 All scientific men were formerly accused of practicing magic. And no wonder, for each said to himself: "I have carried human intelligence as far as it will go, and yet So-and-so has gone further than I. Ergo, he has taken to sorcery."
Baron de la Brède et de Montesquieu, French Philosopher/Author (1689–1755)

76.15 *Linnea*. . . . A plant of Lapland, lowly, insignificant, disregarded, flowering but for a brief space—from Linnaeus who resembles it.
Carl Linnaeus, Swedish Botanist/Naturalist (1707–1778)

76.16 A man [Newton] who, had he flourished in ancient Greece, wouldhave been worshipped as a divinity.
Samuel Johnson, English Lexicographer/Poet/Critic (1709–1784)

76.17 The antechapel where the statue stood
Of Newton with his prism and silent face,
The marble index of a mind forever
Voyaging through strange seas of thought, alone.
William Wordsworth, English Poet (1770–1850)

76.18 The first man of science was he who looked into a thing, not to learn whether it furnished him with food, or shelter, or weapons, or tools, armaments, or playwiths [sic] but who sought to know it for the gratification of knowing.
Samuel Taylor Coleridge, English Poet/Critic (1772–1834)

76.19 To me there never has been a higher source of earthly honor or distinction than that connected with advances in science. I have not possessed enough of the eagle in my character to make a direct flight to the loftiest altitudes in the social world; and I certainly never endeavored to reach those heights by using the creeping powers of the reptile, who in ascending, generally chooses the dirtiest path, because it is the easiest.
Sir Humphrey Davy, English Chemist (1778–1829)

76.20 Newton (that proverb of the mind), alas!
Declared, with all his grand discoveries recent,

That he himself felt only "like a youth
Picking up shells by the great ocean—Truth."
Lord Byron (George Gordon), English Poet/Dramatist (1788–1824)

76.21 O! what a noble heart was here undone,
When Science's self destroyed her favorite son.
[Henry Kirk White, who died from the fatigue of his long research.]
Lord Byron (George Gordon), English Poet/Dramatist (1788–1824)

76.22 The starry Galileo, with his woes.
Lord Byron (George Gordon), English Poet/Dramatist (1788–1824)

76.23 I cannot afford to waste my time making money.
[Reply to an offer of a lecture tour.]
Jean Louis Agassiz, Swiss/American Naturalist/Geologist (1807–1873)

76.24 I fear that the character of my knowledge is from year to year becoming more distinct and scientific; that, in exchange for vistas wide as heaven's scope, I am being narrowed down to the field of the microscope. I see details, not wholes nor the shadow of the whole. I count some parts, and say, "I know."
Henry David Thoreau, American Writer/Naturalist (1817–1862)

76.25 A man [Pasteur himself] whose invincible belief is that science and peace will triumph over ignorance and war, that nations will unite, not to destroy but to build, and that the future will belong to those who will have done the most for suffering humanity. But whether our efforts are or are not favored by life, let us be able to say, when we come near to the great goal, "I have done what I could."
Louis Pasteur, French Chemist/Microbiologist (1822–1895)

76.26 Let me tell you the secret that has led me to my goal. My strength lies in my tenacity.
Louis Pasteur, French Chemist/Microbiologist (1822–1895)

76.27 I have wasted my life.
Louis Pasteur, French Chemist/Microbiologist (1822–1895)

76.28 If at this moment I am not a worn-out, debauched, useless carcass of a man, if it has been or will be my fate to advance the cause of science, if I feel that I have a shadow of a claim on the love of those about me, if in the supreme moment when I looked down into my boy's grave my sorrow was full of submission and without bitterness, it is because these agencies have worked upon me, and not because I have ever cared whether my poor personality shall remain distinct forever from the All from whence it came and whither it goes.

And thus, my dear Kingsley, you will understand what my position is. I may be quite wrong, and in that case I know I shall have to pay the penalty for being wrong. But I can only say with Luther, "Gott helfe mir, ich kann nichts anders [God help me, I cannot do otherwise]."

Thomas Henry Huxley, English Biologist/Evolutionist (1825–1895)

76.29 So little done, so much to do.
[His last words.]

Alexander Graham Bell, Scottish/American Scientist/Inventor (1847–1922)

76.30 Genius is one percent inspiration and ninety-nine percent perspiration.

Thomas Edison, American Inventor (1847–1931)

76.31 If it [the theory of relativity] should prove to be correct, as I expect it will, he [Einstein] will be considered the Copernicus of the twentieth century.

Max Planck, German Physicist (1858–1947)

76.32 Because of the limitations of time, the great majority of us are not able to do justice to our own patients and at the same time take an active part in international affairs, but we should be informed on these matters and we should be aware of the fact that support of health programs beyond the confines of our country serves as an element of great strength in the foreign policy of our Department of State and the production of our own domestic health.

Dr. Charles Mayo, American Physician/Administrator (1865–1939)

76.33 God grant that I may be
More than a nonentity
A little bad, a little good,
Makes but mediocrity.
In daily problems, daily deeds,
Solutions, actions, roses, weeds,
Give me guidance, give me strength
To grow my garden with thy seeds.

Dr. Charles Mayo, American Physician/Administrator (1865–1939)

76.34 I've just been reading some of my early papers and you know, when I'd finished, I said to myself, "Rutherford, my boy, you used to be a damned clever fellow."

Ernest Rutherford, British Physicist (1871–1937)

76.35 The scientist is free, and must be free to ask any question, to doubt any assertion, to seek for any evidence, to correct any errors.

J. Robert Oppenheimer, American Physicist (1904–1967)

76.36 When I find myself in the company of scientists, I feel like a shabby curate who has strayed by mistake into a drawing room full of dukes.

Wystan Hugh Auden, British/American Poet (1907–1973)

76.37 The scientist is indistinguishable from the common man in his sense of evidence, except that the scientist is more careful.

Willard van Orman Quine, American Philosopher (1908–)

76.38 One could not be a successful scientist without realizing that, in contrast to the popular conception supported by newspapers and mothers of scientists, a goodly number of scientists are not only narrow-minded and dull, but also just stupid.

James D. Watson, American Geneticist/Biophysicist (1928–)

76.39 His [Marvin Minsky's] basic interest seemed to be in the workings of the human mind and in making machine models of the mind. Indeed, about that time he and a friend made one of the first electronic machines that could actually teach itself to do something interesting. It monitored electronic "rats" that learned to run mazes. It was being financed by the Navy. On one notable occasion, I remember descending to the basement of Memorial Hall, while Minsky worked on it. It had an illuminated display panel that enabled one to follow the progress of the "rats." Near the machine was a hamster in a cage. When the machine blinked, the hamster would run around its cage happily. Minsky, with his characteristic elfin grin, remarked that on a previous day the Navy contract officer had been down to see the machine. Noting the man's interest in the hamster, Minsky had told him laconically, "The next one we build will look like a bird."

Jeremy Bernstein, American Physicist/Author (1929–)

76.40 To my friends: my work is done. Why wait.
[Suicide note.]

George Eastman, American Chemist (1932)

76.41 As soon as I saw it I decided I was going to spend the rest of my life studying dinosaurs.
[Seeing Rudolph Zallinger's mural of dinosaurs reproduced on the cover of *Life* magazine in spring of 1955, when he was 10 years old.]

Robert Bakker, American Paleontologist/Artist (1945–)

76.42 Paleontology is a very visual inquiry. All paleontologists scribble on napkins at coffee breaks, making sketches to explain their thinking.

Robert Bakker, American Paleontologist/Artist (1945–)

76.43 This is the day we celebrate Bohr
Who gave us the complementarity law
That gives correspondence (as Bohr said before)
That holds in the shell as well as the core
That possesses the compound levels galore
That make up the spectrum
That's due to the modes
That belong to the drop

That looks like the nucleus
That sits in the atom
That Bohr built.
[Composed for the occasion of Bohr's seventieth birthday.]
R.E. Peierls, German Physicist (1955)

76.44 Hell must be isothermal, for otherwise the resident engineers and physical chemists (of which there must be some), could set up a heat engine to run a refrigerator to cool off a portion of their surroundings to any desired temperature.
Henry Albert Bent, British Scientist (1965)

76.45 Pierre Curie, a brilliant scientist, happened to marry a still more brilliant one—Marie, the famous Madame Curie—and is the only great scientist in history who is consistently identified as the husband of someone else.
Isaac Asimov, American Biochemist/Author (1976)

76.46 We then got to Westminster Abbey and, moving about unguided, we found the graves of Newton, Rutherford, Darwin, Faraday, and Maxwell in a cluster.
Isaac Asimov, American Biochemist/Author (1980)

76.47 The mechanical world view is a testimonial to three men: Francis Bacon, René Descartes, and Isaac Newton. After 300 years we are still living off their ideas.
Jeremy Rifkin, American Writer/Environmental Activist (1980)

76.48 As scientists, we need to provide the best information we can. After that, it's a value call.
William J. Schull, American Physicist (1986)

76.49 We may have forgotten that when all is said and done, a good man, as the Greeks say, is a nobler work than a good technologist.
Charles, Prince of Wales, British Royalty (1986)

77.
Sociology

Society is itself a kind of organism, an enormously powerful one, but unfortunately not a very wise one.

ISAAC ASIMOV

77.1 Man perfected by society is the best of all animals; he is the most terrible of all when he lives without law and without justice.
Aristotle, Greek Philosopher (384 B.C.–322 B.C.)

77.2 Do not withdraw from the community.
Hillel, Jewish Rabbi/Scholar/Lawyer (30 B.C.?–9 A.D.)

77.3 It is most true that a natural and secret hatred and aversion towards society, in any man, hath somewhat of the savage beast.
Francis Bacon, English Philosopher/Essayist/Statesman (1561–1626)

77.4 Civilization is simply a series of victories over nature.
William Harvey, English Anatomist/Physician (1578–1657)

77.5 Every individual is continually exerting himself to find out the most advantageous employment for whatever capital he can command. It is his own advantage, indeed, and not that of society, which he has in view. But the study of his own advantage naturally, or rather necessarily, leads him to prefer that employment which is most advantageous to the society.
John Locke, English Philosopher (1632–1704)

77.6 Heaven forming each on other to depend,
A master, or a servant, or a friend,

Bids each on other for assistance call,
Till one man's weakness grows the strength of all.
Alexander Pope, English Poet/Satirist (1688–1744)

77.7 Society is the union of men and not the men themselves.
Baron de la Brède et de Montesquieu, French Philosopher/Author (1689–1755)

77.8 Man is a social animal formed to please in society.
Baron de la Brède et de Montesquieu, French Philosopher/Author (1689–1755)

77.9 Society is as ancient as the world.
François-Marie Arouet de Voltaire, French Author/Philosopher (1694–1778)

77.10 Society itself, which should create
Kindness, destroys what little we had got:
To feel for none is the true social art
Of the world's stoics—men without a heart.
Lord Byron (George Gordon), English Poet/Dramatist (1788–1824)

77.11 Society is a republic. When an individual endeavors to lift himself above his fellows, he is dragged down by the mass, either by means of ridicule or of calumny. No one shall be more virtuous or more intellectually gifted than others. Whoever, by the irresistible force of genius, rises above the common herd is certain to be ostracized by society, which will pursue him with such merciless derision and detraction that at last he will be compelled to retreat into the solitude of his thoughts.
Heinrich Heine, German/Jewish Poet/Author (1797–1856)

77.12 If you would civilize a man, begin with his grandmother.
Victo Hugo, French Poet/Novelist/Dramatist (1802–1885)

77.13 Society does not love its unmaskers.
Ralph Waldo Emerson, American Essayist/Philosopher/Poet (1803–1882)

77.14 Human society is made up of partialities. Each citizen has an interest and a view of his own, which, if followed out to the extreme, would leave no room for any other citizen.
Ralph Waldo Emerson, American Essayist/Philosopher/Poet (1803–1882)

77.15 Society will pardon much to genius and special gifts; but, being in its nature conventional, it loves what is conventional, or what belongs to coming together.

Ralph Waldo Emerson, American Essayist/Philosopher/Poet (1803–1882)

77.16 In this great society wide lying around us, a critical analysis would find very few spontaneous actions. It is almost all custom and gross sense.

Ralph Waldo Emerson, American Essayist/Philosopher/Poet (1803–1882)

77.17 We think our civilization near its meridian, but we are yet only at the cock-crowing and the morning star.

Ralph Waldo Emerson, American Essayist/Philosopher/Poet (1803–1882)

77.18 A sufficient measure of civilization is the influence of good women.

Ralph Waldo Emerson, American Essayist/Philosopher/Poet (1803–1882)

77.19 Nations, like individuals, live and die; but civilization cannot die.

Giuseppe Mazzini, Italian Statesman (1805–1872)

77.20 As the first monogamian family has improved greatly since the commencement of civilization, and very sensibly in our times, it is at least supposable that it is capable of still further improvement until the equality of the sexes is attained.

Lewis Henry Morgan, American Anthropologist (1818–1881)

77.21 Here is the element or power of conduct, of intellect and knowledge, of beauty, and of social life and manners, and all needful to build up a complete human life. . . . We have instincts responding to them all, and requiring them all, and we are perfectly civilized only when all these instincts of our nature—all these elements in our civilization have been adequately recognized and satisfied.

Matthew Arnold, English Poet/Critic (1822–1888)

77.22 The more I see of uncivilized people, the better I think of human nature and the essential differences between civilized and savage men seem to disappear.

Alfred Russel Wallace, English Naturalist/Evolutionist (1823–1913)

77.23 The two poles of social and political philosophy seem necessarily to be organization or anarchy; man's intellect or the forces of nature.

Henry (Brooks) Adams, American Author/Historian (1838–1918)

77.24 A specter is haunting Europe—the specter of Communism. All the Powers of old Europe have entered into a holy alliance to exorcise this specter:

Pope and Czar, Metternich and Guizot, French Radicals and German police-spies.
[Opening paragraph of the *Manifesto of the Communist Party*.]
Karl Marx, German Political Philosopher (1848)

77.25 It is the utmost folly to denounce capital. To do so is to undermine civilization, for capital is the first requisite of every social gain, educational, ecclesiastical, political, or other.
William Graham Sumner, American Sociologist (1840–1910)

77.26 Darwin was as much of an emancipator as was Lincoln.
William Graham Sumner, American Sociologist (1840–1910)

77.27 The meaning of the evolution of culture is no longer a riddle to us. It must present to us the struggle between Eros and Death, between the instincts of life and the instincts of destruction, as it works itself out in the human species.
Sigmund Freud, Austrian Psychiatrist/Psychoanalyst (1856–1939)

77.28 Those who love fairy-tales do not like it when people speak of the innate tendencies in mankind toward aggression, destruction, and, in addition, cruelty.
Sigmund Freud, Austrian Psychiatrist/Psychoanalyst (1856–1939)

77.29 It is only by the influence of individuals who can set an example, whom the masses recognize as their leaders, that they can be induced to submit to the labors and renunciations on which the existence of culture depends.
Sigmund Freud, Austrian Psychiatrist/Psychoanalyst (1856–1939)

77.30 The office of the leisure class in social evolution is to retard movement and to conserve what is obsolescent.
Thorstein Bunde Veblen, American Economist/Sociologist (1857–1929)

77.31 Communism is at once a complete system of proletarian ideology and a new social system. It is different from any other ideological and social system, and is the most complete, progressive, revolutionary, and rational system in human history.
Mao Tse Tung, Chinese Political Leader (1893–1976)

77.32 I have always believed that the character of a society is largely shaped and unified by its great creative works.
Ben Shahn, Lithuanian/American Artist (1898–1969)

77.33 I see the whole of humankind becoming a single, integrated organism. . . . I look upon each of us as I would an individual cell in the organism, each of us playing his or her respective role.
Jonas Salk, American Medical Researcher/Microbiologist (1915–)

77.34 In all systems of human relationship, from the family to the superstate, there is an element of coercion, and this element, whenever it occurs and in whatever degree it is significant or controlling, we conveniently designate as "power."

Charles Wright Mills, American Sociologist (1916–1962)

77.35 Segregation is the offspring of an illicit intercourse between injustice and immorality.

Martin Luther King, American Civil Rights Leader (1929–1968)

78. Space Exploration

We've lost all geographical frontiers on Earth, but new and far larger ones exist at Earth's doorstep.

ISAAC ASIMOV

78.1 Sometimes, for instance, I imagine that I am suspended in the air, and remain there motionless, while the earth turns under me in four-and-twenty hours. I see pass beneath me all these different countenances, some white, others black, others tawny, others olive-colored. At first they wear hats, and then turbans, then heads with long hair, then heads shaven; sometimes towns with steeples, sometimes towns with long spires, which have crescents, sometimes towns with porcelain towers, sometimes extensive countries that have only huts; here wide seas; there frightful deserts; in short, all this infinite variety on the surface of the earth.

Bernard Le Bovier Sieur de Fontenelle, French Philosopher (1657–1757)

78.2 No national sovereignty rules in outer space. Those who venture there go as envoys of the entire human race. Their quest, therefore, must be for all mankind, and what they find should belong to all mankind.

Lyndon Baines Johnson, American President (1908–1973)

78.3 Now is the time to take longer strides—time for a new American enterprise—time for this nation to take a clearly leading role in space achievement, which in many ways may hold the key to our future on earth.

John F. Kennedy, American President (May 25, 1961)

78.4 The exploration and use of outer space, including the moon and other celestial bodies, shall be carried out for the benefit and in the interests of all countries, irrespective of their degree of economic or scientific development, and shall be the province of all mankind.

United Nations Treaty on the Exploration and Use of Space (January 27, 1967)

78.5 One small step for man, one giant leap for mankind.
[Said as he first stepped onto the moon.]

Neil A. Armstrong, American Astronaut/Engineer (1969)

78.6 We came in peace for all mankind.

Plaque left on the moon (July 20, 1969)

78.7 It was an overwhelming sight: there was this structure in flight, with a tremendous amount of apparatus and all sorts of antennas. And huge letters on it spelled out "S.S.S.R."
[On docking with the *Salyut* orbital station while aboard the *Soyuz 10* spacecraft.]

Aleksei S. Yeliseyev, Russian Astronaut (1971)

78.8 The rockets that have made spaceflight possible are an advance that, more than any other technological victory of the twentieth century, was grounded in science fiction. . . .

One thing that no science fiction writer visualized, however, as far as I know, was that the landings on the Moon would be watched by people on Earth by way of television.

Isaac Asimov, American Biochemist/Author (1976)

78.9 It's our UFO to somewhere else!
[On the *Voyager* spacecraft.]

J. Allen Hynek, American Astronomer (1981)

78.10 The information we have so far from the exploration of the planets seems to indicate that the earth is probably the only place in this solar system where there is life.

Cyril Ponnamperuma, American Chemist (1985)

78.11 Those were the sad years in which the joke was that our countdowns ended in "Four, three, two, one, oh shit!"
[On the 1957–1961 years of the U.S. space program.]

Max W. Kraus, American Foreign Services Officer (1985)

78.12 Men do not live in the same place in which they are born. They look for further worlds. . . . And another thing, the spaceman is the only person who can travel without a visa, cross frontiers without a passport, and see the world in ninety minutes!

Georgi Beregovoi, Soviet Astronaut (1985)

78.13 In 20 to 25 years, all the people involved in the early space program will be gone. I have a sense of urgency about producing good paintings [of his experiences] and telling the story, because my time is limited.

Alan Bean, American Astronaut/Painter (1985)

78.14 What we may find out about the material universe may be insignificant by contrast with what we may find out about ourselves. . . . As our distance from that radiant body [Earth] increases, we may discover whether we are fit to inhabit what we have made of it.

Arthur C. Danto, American Philosopher (1985)

78.15 The only choice you've got today is not to go.
[Spoken to officials in the firing room at the Kennedy Space Center in Florida the morning of January 28, 1986, a few hours before liftoff of the shuttle *Challenger*, after noting heavy ice on the launching pad.]

Charles G. Stevenson, American Inspection Team Leader, NASA (1986)

78.16 Uh-oh.
[Last words from *Challenger*, which exploded after taking off, January 28, 1986.]

Michael J. Smith, American Astronaut (1986)

78.17 The first space probe to another star may be a half-mile-wide intelligent aluminum screen door called *Starwisp*. . . . And then it would be off to the stars with a wire-mesh sail, on the wings of microwaves.

Joel Davis, American Journalist (1986)

78.18 Our whole approach to looking for life on Mars has changed in recent years. We are no longer looking for extant life. Rather, we are looking for evidence of extinct life.

Christopher McKay, American Astrophysicist (1986)

78.19 The space station will lie in an orbit several hundred miles above Earth. The black around it will be as black as Margaret Thatcher's Daimler heading for Whitehall, as black as the proverbial lining of the Earl of Hell's waistcoat. It will be cut by hard, bright star points, diamond dust sprinkled everywhere, except where broken by the sphere of the pale, off-white moon.

Robert M. Powers, British Astronomer/Author (1986)

79.
The Stars

Stars look serene, but they are incredibly violent furnaces that occasionally erupt in incredibly violent explosions.

ISAAC ASIMOV

79.1 At this time, too, a fifth star, moving in the opposite direction, was seen to enter the circle of the moon. It is my belief that these portents presaged Theudebald's [King of Reims] death.

Gregory of Tours, Frankish Ecclesiastic/Historian (538–594)

79.2 Then after Easter on the eve of St. Ambrose . . . almost everywhere in this country and almost the whole night, stars in very large numbers were seen to fall from heaven, not by ones or twos, but in such quick succession that they could not be counted.

Anonymous, *The Anglo-Saxon Chronicle* (1094)

79.3 Last year [1572], in the month of November, on the eleventh day of the month, in the evening, after sunset, when, according to my habit, I was contemplating the stars in a clear sky, I noticed that a new and unusual star, surpassing the other stars in brilliancy, was shining almost directly above my head; and since I had, almost from boyhood, known all the stars of the heavens perfectly (there is no great difficulty in attaining that knowledge), it was quite evident to me that there had never before been any star in that place in the sky. . . .

I conclude, therefore, that this star is not some kind of comet or a fiery meteor, whether these be generated beneath the Moon or above the Moon, but that it is a star shining in the firmament itself—one that has never previously been seen before our time, in any age since the beginning of the world.

Tycho Brahe, Danish Astronomer (1546–1601)

79.4 These earthly godfathers of heaven's lights,
That give a name to every fixed star,
Have no more profit of their shining nights
Than those that walk and wot not what they are.
William Shakespeare, English Dramatist/Poet (1564–1616)

79.5 The number of fixed stars which observers have been able to see without artificial powers of sight up to this day can be counted. It is therefore decidedly a great feat to add to their number, and to set distinctly before the eyes other stars in myriads, which have never been seen before, and which surpass the old, previously known stars in number more than ten times.
Galileo Galilei, Italian Astronomer/Physicist (1564–1642)

79.6 The next object which I have observed is the essence or substance of the Milky Way. By the aid of a telescope anyone may behold this in a manner which so distinctly appeals to the senses that all the disputes which have tormented philosophers through so many ages are exploded at once by the irrefragable evidence of our eyes, and we are freed from wordy disputes upon this subject, for the Galaxy is nothing else but a mass of innumerable stars planted together in clusters.
Galileo Galilei, Italian Astronomer/Physicist (1564–1642)

79.7 So when, by various turns of the Celestial Dance,
In many thousand years,
A Star, so long unknown, appears,
Tho' Heaven itself more beauteous by it grow,
It troubles and alarms the World below,
Does to the Wise a Star, to Fools a Meteor show.
Abraham Cowley, English Poet (1618–1667)

79.8 Spong and I had also several fine discourses upon the globes this afternoon, particularly why the fixed stars do not rise and set at the same hour all the year long, which he could not demonstrate nor I neither.
Samuel Pepys, English Diarist (1633-1703)

79.9 When our philosophers are asked what is the use of these countless myriads of fixed stars, of which a small part would be sufficient to do what they all do, they coolly tell us that they are made to give delight to their eyes.
Bernard Le Bovier Sieur de Fontenelle, French Philosopher (1657–1757)

79.10 The stars hang bright above,
Silent, as if they watch'd the sleeping earth.
Samuel Taylor Coleridge, English Poet/Critic (1772–1834)

79.11 When I consider how, after sunset, the stars come out gradually in troups from behind the hills and woods, I confess that I could not have contrived a more curious and inspiring night.
Henry David Thoreau, American Writer/Naturalist (1817–1862)

79.12 I just looked up at a fine twinkling star and thought that a voyager whom I know, now many a days' sail from this coast, might possibly be looking up at that same star with me. The stars are the apexes of what triangles!

Henry David Thoreau, American Writer/Naturalist (1817–1862)

79.13 The conditions of the earth's core are starlike. From their study can physicists of the future tell us something more of the true nature of the stars?

Reginald A. Daly, American Geologist/Geophysicist (1871–1957)

79.14 They [brown dwarf stars] could last the age of the universe. Compare that to the sun, which is only 5 billion years old, or about one-third the age of the universe. After another 5 billion years, the sun will use up its nuclear fuel and at least partially blow up and leave behind a white dwarf. Whereas a brown dwarf just sort of sits there and cools, and sits there and cools a little more, and will just stay there forever, cooling gradually.

Paul C. Joss, American Astrophysicist (1986)

79.15 Supernovas are the final act in the lives of many stars. They are also the only plausible place for the synthesis of the heavier chemical elements. The explosions eject these elements, plus those made in previous stages of the star's life cycle, into space, where they are available for recycling into new generations of stars and planets.

Dietrick E. Thomsen, American Science Journalist (1986)

79.16 The dark clouds in the constellation Orion are one of the astronomer's favorite places to look for stars that are just beginning to form.

Dietrick E. Thomsen, American Science Journalist (1986)

79.17 It's as if God took a piece of rope and bent it there.
[Concerning bright luminous intergalactic arcs discovered by Roger Lynds of Kitt Peak National Observatory at the National Optical Astronomy Observatories, Tucson.]

Vahe Petrosian, Science Researcher (1987)

80. Subjective Sciences

The wish to believe, even against evidence, fuels all the pseudosciences from astrology to creationism.

ISAAC ASIMOV

80.1 Astrology is a sickness, not a science. . . . It is a tree under the shade of which all sorts of superstitions thrive.

Maimonides, Jewish Philosopher/Physician (1135–1204)

80.2 The alchemists call in many varieties out of astrology, auricular traditions, and feigned testimonies.

Francis Bacon, English Philosopher/Essayist/Statesman (1561–1626)

80.3 I was ever of the opinion that the philosopher's stone, and an holy war, were but the rendezvous of cracked brains, that wore their feather in their heads.

Francis Bacon, English Philosopher/Essayist/Statesman (1561–1626)

80.4 That there should be more Species of intelligent Creatures above us, than there are of sensible and material below us, is probable to me from hence; That in all the visible corporeal World, we see no Chasms, or no Gaps.

Joseph Addison, English Essayist/Poet/Politician (1672–1719)

80.5 It is the characteristic of true science, to discern the impassable, but not very obvious, limits which divide the province of reason from that of speculation. Such knowledge comes tardily. How many ages have rolled away in which powers, that, rightly directed, might have revealed the great laws of nature, have been wasted in brilliant, but barren, reveries on alchemy and astrology?

William Hickling Prescott, American Historian (1796–1859)

80.6 In the realm of science all attempts to find any evidence of supernatural beings, of metaphysical conceptions, as God, immortality, infinity, etc., thus far have failed, and if we are honest we must confess that in science there exists no God, no immortality, no soul or mind as distinct from the body.

Charles P. Steinmetz, German/American Engineer/Inventor (1865–1923)

80.7 Undeterred by poverty, failure, domestic tragedy, and persecution, but sustained by his mystical belief in an attainable mathematical harmony and perfection of nature, Kepler persisted for fifteen years before finding the simple regularity [of planetary orbits] he sought. . . . What stimulated Kepler to keep slaving all those fifteen years? An utter absurdity. In addition to his faith in the mathematical perfectibility of astronomy, Kepler also believed wholeheartedly in astrology. This was nothing against him. For a scientist of Kepler's generation astrology was as respectable scientifically and mathematically as the quantum theory or relativity is to theoretical physicists today. Nonsense now, astrology was not nonsense in the sixteenth century.

Eric Temple Bell, American Mathematician/Author (1937)

80.8 We are about to move into the Aquarian age of clearer thinking. Astrology and witchcraft both have a contribution to make to the new age, and it behooves the practitioners of both to realize their responsibilities and obligations to the science and the religion.

Sybil Leek, English/American Writer/Witch (1968)

80.9 When there is publicity about [UFO] sightings that turn out to be explainable, the percentage of unexplained sightings goes up, suggesting that these, too, are caused by something in people's psychology rather than by something that is actually out there. The UFO evidence forms no coherent residue; it never gets better.

Francis Crick, British Biophysicist/Geneticist (1984)

80.10 There is consciousness without content, oceanic consciousness. It could be that the techniques of yoga and Zen have manipulated the neurochemistry of systems related to the limbic brain.

Karl Pribram, Austrian/American Neurophysiologist (1984)

80.11 On vitamin K, I have experienced states in which I can contact the creators of the universe, as well as the *local* creative controllers—the Earth Coincidence Control Office, or ECCO. They're the guys who run earth and who program us, though we're not aware of it. I asked them, "What's your major program?" They answered, "To make you guys evolve to the next levels, to teach you, to kick you in the pants when necessary."

John Lilly, American Neurophysiologist/Spiritualist (1984)

80.12 The greatest scientists have always looked on scientific materialism as a kind of religion, as a mythology. They are impelled by a great desire to explore mystery, to celebrate mystery in the universe, to open it up, to read the stars, to find the deeper meaning.

Edward O. Wilson, American Entomologist/Sociobiologist (1984)

80.13 One of the most interesting exhibits will be the only Bigfoot hair and blood samples ever authenticated by American scientists. The samples weren't exactly identified as coming from Bigfoot but it was proved that the hair did not come from any of eighty-six North American mammals. The closest match was a gorilla, but it was not exact.

Erik Beckjord, American Curator of the Cryptozoological Museum (1986)

81.
The Sun

The Sun is all in all to us, the center from which all arises, but look wider, and it is only one of countless billions.

ISAAC ASIMOV

81.1 The heavens declare the glory of God; and the firmament sheweth his handiwork. . . . In them hath he set a tabernacle for the sun, which is as a bridegroom coming out of his chamber, and rejoiceth as a strong man to run a race. His going forth is from the end of the heaven, and his circuit unto the ends of it: and there is nothing his from the heat thereof.

The Bible (circa 725 B.C.)

81.2 The sun . . . is a body of great size and power, the ruler, not only of the seasons and of the different climates, but also of the stars themselves and of the heavens. When we consider his operations, we must regard him as the life, or rather the mind of the universe, the chief regulator and the God of nature; he also lends his light to the other stars. He is the most illustrious and excellent, beholding all things and hearing all things.

Pliny the Elder (Gaius Plinius Secundus), Roman Naturalist/Historian (23–79)

81.3 Before the great plague [in 571 A.D.] which ravaged Auvergne prodigies terrified the people of the region in the same way. On a number of occasions three or four great shining lights appeared round the sun, and these the country folk also called suns. . . . Once, on the first day of October, the sun was in eclipse, so that less than a quarter of it continued to shine, and the rest was so dark and discolored that you would have said that it was made of sackcloth.

Gregory of Tours, Frankish Ecclesiastic/Historian (538–594)

81.4 Since Britain lies far north toward the pole, the nights are short in summer, and at midnight it is hard to tell whether the evening twilight still lingers or whether dawn is approaching, since the sun at night passes not far below the earth in its journey round the north back to the east. Consequently the days are long in summer, as are the nights in winter when the sun withdraws into African regions.

Bede, English Monk/Scholar (673?–735)

81.5 In this year the first day of Whitsuntide was on 5 June, and on the following Tuesday at noon there appeared four intersecting halos around the sun, white in color, and looking as if they had been painted. All who saw it were astonished, for they did not remember seeing anything like it before.

Anonymous, *The Anglo-Saxon Chronicle* (1094)

81.6 From every side the sun shot forth the light of day, so that he had with his bright rays chased Capricorn from mid-heaven.

Dante Alighieri, Italian Poet (1265–1321)

81.7 In the middle of everything is the sun. For in this most beautiful temple, who would place this lamp in another or better position than that from which it can light up the whole thing at the same time? For, the sun is not inappropriately called by some the lantern of the universe, by others, its mind, and, its ruler by others still. . . . Thus indeed, as though seated on a royal throne, the sun rules the family of planets revolving around it.

Nicholas Copernicus, Polish Astronomer (1473–1543)

81.8 The sun alone appears, by virtue of his dignity and power, suited for this motive duty (of moving the planets) and worthy to become the home of God himself.

Johannes Kepler, German Astronomer/Mathematician (1571–1630)

81.9 The glorious lamp of heaven, the sun.

Robert Herrick, English Poet (1591–1674)

81.10 Why there is one Body in our System qualified to give Light and Heat to all the rest, I know no Reason, but because the Author of the System thought it convenient.

Isaac Newton, English Physicist/Mathematician (1642–1727)

81.11 How could our eyes see the sun, unless they are sunlike themselves?

Johann Wolfgang von Goethe, German Poet/Dramatist/Novelist (1749–1832)

81.12 Our sun, by the way . . . may become a white dwarf some day but apparently will never become a supernova.

Isaac Asimov, American Biochemist/Author (1965)

81.13 The sun rises. . . .

I mean, just for a start, that the sun shines. That four million tons of matter are destroyed at its surface every second in a raging nuclear storm which buffets us here on the fringes of the solar atmosphere. And all the time, the mass of our star is being reduced and its volume and density are changing.

Lyall Watson, American Biologist (1979)

81.14 The sun's surface does not rotate rigidly, uniformly at the same speed; latitudes nearer the equator rotate faster than the polar ones. However, observations a few years ago showed that, contrary to expectations, the rotation rate does not increase smoothly from poles to equator. Instead, there is a small wiggle superimposed on the smooth change, so that alternately a slice of the surface will be going slightly faster and the slice next to it slightly slower than they might if the change in rotation speed were smooth.

Dietrick E. Thomsen, American Science Journalist (1987)

82.
Taxonomy

The card-player begins by arranging his hand for maximum sense. Scientists do the same with the facts they gather.

ISAAC ASIMOV

82.1 When I first read Plato and came upon this gradation of beings which rises from the lightest atom to the Supreme Being, I was struck with admiration. But when I looked at it more closely, the great phantom vanished. . . . At first the imagination takes a pleasure in seeing the imperceptible transition from inanimate to organic matter, from plants to zoophytes, from these to animals, from these to genii, from these genii endued with a small aerial body to immaterial substances, and finally angels. This hierarchy pleases those good folk who fancy they see it in the Pope and his cardinals, followed by archbishops and bishops, after whom come the rectors, the vicars, the simple curates, the deacons, the subdeacons, then the monks and the line is ended by the Capuchins.

François-Marie Arouet de Voltaire, French Author/Philosopher (1694–1778)

82.2 If the point were once gained that among animals and vegetables, there had been, I do not say several species, but even a single one, which had been produced in the course of direct descent from other species . . . then no further limit could be set to the power of nature, and we should not be wrong in supposing, with sufficient time, that she could have developed all other organic forms from one primordial type.

Georges-Louis Leclerc, Comte de Buffon, French Naturalist (1707–1788)

82.3 It is the genus that gives the characters, and not the characters that make the genus.

Carl Linnaeus, Swedish Botanist/Naturalist (1707–1778)

82.4 Nature has not arranged her productions on a single and direct line. They branch at every step, and in every direction, and he who attempts to reduce them into departments is left to do it by the lines of his own fancy.

Thomas Jefferson, American President/Author (1743–1826)

82.5 All known living bodies are sharply divided into two special kingdoms, based upon the essential differences which distinguish animals from plants, and in spite of what has been said, I am convinced that these two kingdoms do not really merge into one another at any point.

Jean-Baptiste de Monet, Chevalier de Lamarck, French Naturalist (1744–1829)

82.6 In the real changes which animals undergo during their embryonic growth, in those external transformations as well as in those structural modifications within the body, we have a natural scale to measure the degree or the gradation of those full grown animals which corresponds in their external form and in their structure, to those various degrees in the metamorphoses of animals, as illustrated by embryonic changes, a real foundation for zoological classification.

Jean Louis Agassiz, Swiss/American Naturalist/Geologist (1807–1873)

82.7 The natural system [of birds] may, perhaps, be most truly compared to an irregularly branching tree, or rather to an assemblage of detached trees and shrubs of various sizes and modes of growth. And as we show the form of a tree by sketching it on paper, or by drawing its individual branches and leaves, so may the natural system be drawn on a map, and its several parts shown in greater detail in a series of maps.

Hugh Strickland, English Ornithologist (1811–1853)

82.8 Every species has come into existence coincident both in space and time with a preexisting closely allied species.

Alfred Russel Wallace, English Naturalist/Evolutionist (1823–1913)

82.9 There may be as many classifications of any series of natural, or of other, bodies, as they have properties or relations to one another, or to other things; or, again, as there are modes in which they may be regarded by the mind: so that, with respect to such classifications as we are here concerned with, it might be more proper to speak of *a* classification than of *the* classification of the animal kingdom.

Thomas Henry Huxley, English Biologist/Evolutionist (1825–1895)

82.10 It is a remarkable fact that some well-separated phylogenetic types, thanks to certain biological influences, become so similar in their external appearance that not only is the inexperienced eye led into error, but biologists may inexactly appreciate their real affinities.

Alfred Giard, French Zoologist/Evolutionist (1846–1908)

82.11 The concepts and methods on which the classification of hominid taxa is based do not differ in principle from those used for other zoological taxa. Indeed, the classification of living human populations or of samples of fossil hominids is a branch of animal taxonomy.

Ernst Mayr, American Zoologist (1963)

82.12 As new areas of the world came into view through exploration, the number of identified species of animals and plants grew astronomically. By 1800 it had reached 70,000. Today more than 1.25 million different species, two-thirds animal and one-third plant, are known, and no biologist supposes that the count is complete.

Isaac Asimov, American Biochemist/Author (1965)

82.13 Scientific order is logical, natural, and *forbidding;* while alphabetical order is illogical, conventional, and *inviting.*

Hugh Davidson, American Educator (1972)

82.14 Taxonomy is often regarded as the dullest of subjects, fit only for mindless ordering and sometimes denigrated within science as mere "stamp collecting" (a designation that this former philatelist deeply resents). If systems of classification were neutral hat racks for hanging the facts of the world, this disdain might be justified. But classifications both reflect and direct our thinking. The way we order represents the way we think. Historical changes in classification are the fossilized indicators of conceptual revolutions.

Stephen Jay Gould, American Biologist/Author (1983)

83. Technology

Science in the service of humanity is technology, but lack of wisdom may make the service harmful.

ISAAC ASIMOV

83.1 It is well to observe the force and virtue and consequence of discoveries, and these are to be seen nowhere more conspicuously than in printing, gunpowder, and the magnet. For these three have changed the whole face and state of things throughout the world.
Francis Bacon, English Philosopher/Essayist/Statesman (1561–1626)

83.2 I sell here, sir, what all the world desires to have—power.
[About the improved steam engine invented by himself and James Watt.]
Matthew Boulton, English Engineer/Inventor (1728–1809)

83.3 It would be a strange anomaly if the science of the nation were declining whilst the general intelligence and prosperity increase.
William Vernon Harcourt, English Founder of Science Societies (1789–1871)

83.4 Steam is no stronger now than it was a hundred years ago, but it is put to better use.
Ralph Waldo Emerson, American Essayist/Philosopher/Poet (1803–1882)

83.5 Blessings on Science, and her handmaid Steam!
They make Utopia only half a dream;
And show the fervent, of capacious souls,
Who watch the ball of Progress as it rolls,

That all as yet completed, or begun,
Is but the dawning that precedes the sun.
Charles Mackay, Scottish Poet (1814–1889)

83.6 Each workman would receive two or three important parts and would affix them together and pass them on to the next who would add a part and pass the growing article to another who would do the same . . . until the complete arm is put together.
[Describing the production of Colt's revolving chamber gun, the "equalizer," on the first factory assembly line.]
Samuel Colt, American Inventor/Industrialist (1814–1862)

83.7 No category of sciences exists to which one could give the name of applied sciences. There are science and the applications of science, linked together as fruit is to the tree that has borne it.
Louis Pasteur, French Chemist/Microbiologist (1822–1895)

83.8 By research in pure science I mean research made without any idea of application to industrial matters but solely with the view of extending our knowledge of the Laws of Nature. I will give just one example of the "utility" of this kind of research, one that has been brought into great prominence by the War—I mean the use of x-rays in surgery. Now, how was this method discovered? It was not the result of a research in applied science starting to find an improved method of locating bullet wounds. This might have led to improved probes, but we cannot imagine it leading to the discovery of x-rays.
Sir Joseph John Thomson, English Physicist (1856–1940)

83.9 The path of civilization is paved with tin cans.
Elbert Hubbard, American Editor/Educator (1856–1915)

83.10 One machine can do the work of fifty ordinary men. No machine can do the work of one extraordinary man.
Elbert Hubbard, American Editor/Educator (1856–1915)

83.11 [Edison] definitely ended the distinction between the theoretical man of science and the practical man of science, so that today we think of scientific discoveries in connection with their possible present or future application to the needs of man. He took the old rule-of-thumb methods out of industry and substituted exact scientific knowledge, while, on the other hand, he directed scientific research into useful channels.
Henry Ford, American Industrialist/Auto Maker (1863–1947)

83.12 These expert men, technologists, engineers, or whatever name may best suit them, make up the indispensable General staff of the industrial system; and without their immediate and unremitting guidance and correction the industrial system will not work. It is a mechanically organized structure of technical processes designed, installed, and conducted by these production engineers. Without them and their constant attention the industrial equipment, the mechanical appliances of industry, will foot up to just so much junk.

Thorstein Bunde Veblen, American Economist/Sociologist (1857–1929)

83.13 One must not forget that progress in fundamental science inevitably leads in many cases to a revolution in technology and the daily life of all people. A country which in our epoch ignores this truth dooms itself sooner or later to lagging behind in science and technology.

Leonid I. Sedov, Russian Aerospace Administrator (1971)

83.14 Society has reached a stage of development where the stresses and strains produced by its own speed of technological advance are not only overtaking man's powers of adaptability—both physical and mental—but are endangering his very survival.

Joint Statement of Swiss Ecologists, Swiss Science Journal, *Experientia* (1971)

83.15 The Indian interest in creating standards of care, enforcing them or even extending them, and of protecting its citizens from ill use is significantly stronger than the local interest in deterring multinationals from exporting allegedly dangerous technologies.
[From his ruling that the Indian Government should try their own suit against Union Carbide concerning the death of over 2,000 persons from a poisonous gas leak.]

John F. Keenan, American Judge (1984)

84.
Theory

Facts are a heap of bricks and timber.
It is only a successful theory that can
convert the heap into a stately mansion.

ISAAC ASIMOV

84.1 Nothing is so firmly believed as what we least know.
Michel de Montaigne, French Essayist (1533–1592)

84.2 The confirmation of theories relies on the compact adaption of their parts, by which, like those of an arch or dome, they mutually sustain each other, and form a coherent whole.
Francis Bacon, English Philosopher/Essayist/Statesman (1561–1626)

84.3 We are under obligation to the ancients for having exhausted all the false theories that could be formed.
Bernard Le Bovier Sieur de Fontenelle, French Philosopher (1657–1757)

84.4 It is with theories as with wells: you may see to the bottom of the deepest if there be any water there, while another shall pass for wondrous profound when 'tis merely shallow, dark, and empty.
Jonathan Swift, Irish Satirist/Clergyman (1667–1745)

84.5 He [Samuel Johnson] bid me always remember this, that after a system is well settled upon positive evidence, a few objections ought not to shake it. "The human mind is so limited that it cannot take in all parts of a subject; so that there may be objections raised against anything. There are objections against a *plenum,* and objections against a *vacuum.* Yet one of them must certainly be true."

[*Plenum* refers to the theory of space being full of matter; *vacuum* refers to the theory that parts of space are empty of matter.]

James Boswell, Scottish Lawyer/Author (1740–1795)

84.6 The moment a person forms a theory, his imagination sees, in every object, only the traits which favor that theory.

Thomas Jefferson, American President/Author (1743–1826)

84.7 It is always better to have no ideas than false ones; to believe nothing than to believe what is wrong.

Thomas Jefferson, American President/Author (1743–1826)

84.8 The World little knows how many thoughts and theories which have passed through the mind of a scientific investigator and have been crushed in silence and secrecy by his own severe criticism and adverse examinations; that in the most successful instances not a tenth of the suggestions, the hopes, the wishes, the preliminary conclusions have been realized.

Michael Faraday, English Chemist/Physicist (1791–1867)

84.9 This word [theory] is employed by English writers in a very loose and improper sense. It is with them usually convertible into *hypothesis,* and *hypothesis* is commonly used as another term for *conjecture*. The terms *theory* and *theoretical* are properly used in opposition to the terms *practice* and *practical*. In this sense they were exclusively employed by the ancients; and in this sense they are almost exclusively employed by the continental philosophers.

William Rowan Hamilton, Irish Mathematician (1805–1865)

84.10 Men who have excessive faith in their theories or ideas are not only ill-prepared for making discoveries; they also make poor observations.

Claude Bernard, French Physiologist (1813–1878)

84.11 Before any great scientific principle receives distinct enunciation by individuals, it dwells more or less clearly in the general scientific mind. The intellectual plateau is already high, and our discoverers are those who, like peaks above the plateau, rise a little above the general level of thought at the time.

John Tyndall, English Physicist (1820–1893)

84.12 It is the customary fate of new truths to begin as heresies and to end as superstitions.

Thomas Henry Huxley, English Biologist/Evolutionist (1825–1895)

84.13 Physical concepts are free creations of the human mind, and are not, however it may seem, uniquely determined by the external world. In our endeavor to understand reality we are somewhat like a man trying to understand the mechanism of a closed watch. He sees the face and the moving hands, even hears its ticking, but he has no way of opening the case. If he is hands, even hears its ticking, but he has no way of opening the case. If he is ingenious, he may form some picture of a mechanism which could be responsible for all the things he observes, but he may never be quite sure his picture is the only one which could explain his observations. He will never be able to compare his picture with the real mechanism and he cannot even imagine the possibility of the meaning of such a comparison.

Albert Einstein, German/American Physicist (1879–1955)

84.14 Creating a new theory is not like destroying an old barn and erecting a skyscraper in its place. It is rather like climbing a mountain, gaining new and wider views, discovering unexpected connections between our starting point and its rich environment. But the point from which we started out still exists and can be seen, although it appears smaller and forms a tiny part of our broad view gained by the mastery of the obstacles on our adventurous way up.

Albert Einstein, German/American Physicist (1879–1955)

84.15 We are all agreed that your theory is crazy. The question that divides us is whether it is crazy enough to have a chance of being correct. My own feeling is that it is not crazy.
[Commenting as spokesperson for a group of physicists reviewing a theory of subnuclear particles proposed by Werner Heisenberg and Wolfgang Pauli.]

Niels Henrik David Bohr, Danish Physicist (1885–1962)

84.16 The farther an experiment is from theory, the closer it is to the Nobel Prize.

Irene Joliot-Curie, French Chemist (1897–1956)

84.17 When a discovery has finally won tardy recognition it is usually found to have been anticipated, often with cogent reasons and in great detail. Darwinism, for instance, may be traced back through the ages to Heraclitus and Anaximander.

F.C.S. Schiller, English Philosopher (1917)

84.18 The mind likes a strange idea as little as the body likes a strange protein and resists it with similar energy. It would not perhaps be too fanciful to say that a new idea is the most quickly acting antigen known to science. If we watch ourselves honestly, we shall often find that we have begun to argue against a new idea even before it has been completely stated.

Wilfred Trotter, English Philosopher/Scientist (1941)

84.19 Newton's passage from a falling apple to a falling moon was an act of the prepared imagination. Out of the facts of chemistry the constructive imagination of Dalton formed the atomic theory. Davy was richly endowed with the imaginative faculty, while with Faraday its exercise was incessant, preceding, accompanying, and guiding all his experiments. His strength and fertility as a discoverer are to be referred in great part to the stimulus of the imagination.

Rosamund E.M. Harding, English Scientist (1942)

84.20 In 1900 however, he [Planck] worked out the revolutionary quantum theory, a towering achievement which extended and improved the basic concepts of physics. It was so revolutionary, in fact, that almost no physicist, including Planck himself, could bring himself to accept it. (Planck later said that the only way a revolutionary theory could be accepted was to wait until all the old scientists had died.)
Isaac Asimov, American Biochemist/Author (1976)

84.21 Facts are not pure unsullied bits of information; culture also influences what we see and how we see it. Theories, moreover, are not inexorable inductions from facts. The most creative theories are often imaginative visions imposed upon facts; the source of imagination is also strongly cultural.
Stephen Jay Gould, American Biologist/Author (1981)

84.22 A lesson to be learned from this puzzle is that it is not necessarily the pieces which seem to fit satisfactorily that decide whether or not the picture to date is correct, it is the pieces which don't fit that really decide the issue.
Peter Warlow, British Physicist/Author/Lecturer (1982)

84.23 In all scientific fields, theory is frequently more important than experimental data. Scientists are generally reluctant to accept the existence of a phenomenon when they do not know how to explain it. On the other hand, they will often accept a theory that is especially plausible before there exists any data to support it.
Richard Morris, American Physicist/Author (1983)

84.24 You can start a fire just by rubbing two dry theories together.
Robert M. Powers, British Astronomer/Author (1986)

85.
The Universe

Since the Universe is defined as including all that exists, it is useless to ask what lies beyond it.

ISAAC ASIMOV

85.1 Chaos was born first and after her came Gaia
the broad-breasted, the firm seat of all
the immortals who hold the peaks of snowy Olympus,
and the misty Tartaros in the depths of broad-pathed earth
and Eros, the fairest of the deathless gods;

Chaos gave birth to Ether and Day.
Gaia now first gave birth to starry Ouranos,
her match in size, to encompass all of her,
and be the firm seat of all the blessed gods.

[Gaie = Earth; Ouranos = sky; Tartaros = the darkest part of the nether world; Erebos = darkness.]

Hesiod, Greek Poet (circa 800 B.C.)

85.2 In the beginning God created the heavens and the earth. Now the earth was a formless void, there was darkness over the deep, and God's spirit hovered over the water. . . . God said, "Let there be a vault in the waters to divide the waters in two." And so it was. God made the vault, and it divided the waters above the vault from the waters under the vault. God called the vault "heaven."

The Bible **(circa 725 B.C.)**

85.3 Before the ocean was, or earth, or heaven,
Nature was all alike, a shapelessness,

Chaos, so-called, all rude and lumpy matter,
Nothing but bulk, inert, in whose confusion
Discordant atoms warred. . . .
Till God, or kindlier Nature,
Settled all argument, and separated
Heaven from earth, water from land, our air
From the high stratosphere, a liberation
So things evolved, and out of blind confusion
Found each its place, bound in eternal order.

Ovid (Publius Ovidus Nasso), Roman Poet (43 B.C.–17 A.D.?)

85.4 But if the heavens are moved by a daily movement, it is necessary to assume in the principal bodies of the universe and in the heavens two ways of movement which are contrary to each other: one from east to west and the other from west to east, as has often been said. And with this, it is proper to assume an excessively great speed, for anyone who reckons and considers well the height of distance of the heavens and the magnitude of these and of their circuit, if such a circuit were made in a day, could not imagine or conceive how marvelously and excessively swift would be the movement of the heavens, and how unbelievable and unthinkable.

Nicole Oresme, French Philosopher/Clergyman (1323–1382)

85.5 The universe, then, has no circumference, for, if it had a center and a circumference, it would thus have in itself its beginning and its end, and the universe itself would be terminated by relation to something else; there would be outside the universe another thing and a place— but all this contains no truth.

Nicholas of Cusa, German Cardinal/Mathematician/Statesman (1401–1464)

85.6 Near the sun is the center of the universe.

Nicholas Copernicus, Polish Astronomer (1473–1543)

85.7 I have said that, in my opinion, all was chaos . . . and out of that bulk a mass formed—just as cheese is made out of milk—and worms appeared in it, and these were the angels. The most holy majesty decreed that these should be God, he too having been created out of that mass at the same time.
[Taken from his testimony before the Inquisition, which sentenced him to burning at the stake.]

Domenico Scandella, Italian Miller (1532–1599)

85.8 This very sun, this very moon, these stars, this very order and revolution of the universe, is the same which your ancestors enjoyed, and which will be the admiration of your posterity.

Michel de Montaigne, French Essayist (1533–1592)

85.9 The first question concerning the Celestial Bodies is whether there be a system, that is whether the world or universe compose together one globe, with a center, or whether the particular globes of earth and stars be scattered dispersedly, each on its own roots, without any system or common center.

Francis Bacon, English Philosopher/Essayist/Statesman (1561–1626)

85.10 It is no small comfort when I reflect that we should not so much marvel at the vast and almost infinite breadth of the most distant heavens but much more at the smallness of us manikins and the smallness of this our tiny ball of earth and also of all the planets.

Johannes Kepler, German Astronomer/Mathematician (1571–1630)

85.11 Moreover, I began to contemplate the vast magnitude of the universe and what proportion this poor globe of earth might bear with it, for if those numberless bodies which stick in the vast roof of heaven, though they appear to us but as spangles, be some of them thousands of times bigger than the earth . . . surely the astronomers had reason to term this sphere an indivisible point and a thing of no dimension at all being compared to the whole world.

James Howell, Welsh Author (1594–1666)

85.12 Give me extension and motion, and I will construct the Universe.

René Descartes, French Mathematician/Philosopher (1596–1650)

85.13 The center of the universe is the most noble place in the world, for it is everywhere distant from the extremes and maintains the middle position.

Giovanni Baptista Riccoli, Italian Philosopher (1598–1671)

85.14 A wonder it must be that there should be any man found so stupid as to persuade himself that this most beautiful world could be produced by the fortuitous concourse of atoms.

John Ray, English Naturalist (1627–1705)

85.15 When the heavens were a little blue arch, stuck with stars, methought the Universe was too straight and close: I was almost stifled for want of air: but now it is enlarged in height and breadth, and a thousand vortices taken in. I begin to breathe with more freedom, and I think the Universe to be incomparably more magnificent than it was before.

Bernard Le Bovier Sieur de Fontenelle, French Philosopher (1657–1757)

85.16 This great work, always more amazing in proportion as it is better known, raises in us so grand an idea of its Maker, that we find our mind overwhelmed with feelings of wonder and adoration.

Bernard Le Bovier Sieur de Fontenelle, French Philosopher (1657–1757)

85.17 That the universe was formed by a fortuitous concourse of atoms, I will no more believe than that the accidental jumbling of the alphabet would fall into a most ingenious treatise of philosophy.

Jonathan Swift, Irish Satirist/Clergyman (1667–1745)

85.18 Man is born not to solve the problems of the universe, but to find out where the problem begins, and then to restrain himself within the limits of the comprehensible.

Johann Wolfgang von Goethe, German Poet/Dramatist/Novelist (1749–1832)

85.19 Taken as a whole, the universe is absurd.

Walter Savage Landor, English Poet/Author (1775–1864)

85.20 I don't pretend to understand the Universe—it's a great deal bigger than I am. . . . People ought to be modester.

Thomas Carlyle, Scottish Historian/Philosopher (1795–1881)

85.21 The universe is a disymmetrical whole.

Louis Pasteur, French Chemist/Microbiologist (1822–1895)

85.22 A man said to the universe, "Sir, I exist." "However," replied the universe, "the fact has not created in me a sense of obligation."

Stephen Crane, American Writer (1871–1900)

85.23 Science is a great game. It is inspiring and refreshing. The playing field is the universe itself.

Isidor Isaac Rabi, Austrian/American Physicist (1898–1963)

85.24 As we push ever more deeply into the universe, probing its secrets, discovering its way, we must also constantly try to learn to cooperate across the frontiers that really divide earth's surface.

Lyndon Baines Johnson, American President (1908–1973)

85.25 If there is life elsewhere in the universe, chemically speaking, it would be very similar to what we have on earth.

Cyril Ponnamperuma, American Chemist (1925–)

85.26 The universe is in the business of making life.

Cyril Ponnamperuma, American Chemist (1925–)

85.27 Consciousness may be associated with all quantum mechanical processes . . . since everything that occurs is ultimately the result of one or more quantum mechanical events, the universe is "inhabited" by an almost unlimited number of rather discrete, conscious, usually nonthinking entities that are responsible for the detailed working of the universe.

Evan H. Walker, American Physicist (1970)

85.28 It is tempting to wonder if our present universe, large as it is and complex though it seems, might not be merely the result of a very slight random increase in order over a very small portion of an unbelievably colossal universe which is virtually entirely in heat-death.

Perhaps we are merely sliding down a gentle ripple that has been set up, accidently and very temporarily, in a quiet pond, and it is only the limitation of our own infinitesimal range of viewpoint in space and time that makes it seem to ourselves that we are hurtling down a cosmic waterfall of increasing entropy, a waterfall of colossal size and duration.

Isaac Asimov, American Biochemist/Author (1976)

85.29 The big bang, that most cataclysmic of all events, now seems to have been a strangely gentle explosion, not so much a violent blowout as a finely orchestrated event—more like the steady inflation of a football than the burst of a bomb. It begins to look as though the cosmos might just be breathing out, and when the current exhalation ends, perhaps after another ten geos, it will pause and then inhale.

Lyall Watson, American Biologist (1979)

85.30 Do you know about the Eleventh Commandment? It says, "Thou shalt not bore God, or he will destroy your universe."

John Lilly, American Neurophysiologist/Spiritualist (1984)

85.31 If the universe is measurably curved today, cosmologists must accept the miraculous fact that this is so for the first time in the 10^{10}–year history of the universe; if it had been measurably nonflat at much earlier times, it would be much more obviously curved today than it is. This line of reasoning suggests that the observable universe is essentially exactly flat: that it contains precisely the critical density of mass.

Lawrence M. Krauss, American Physicist/Astronomer (1986)

85.32 Astrophysicists closing in on the grand structure of matter and emptiness in the universe are ruling out the meatball theory, challenging the soap bubble theory, and putting forward what may be the strongest theory of all: that the cosmos is organized like a sponge.

James Gleick, American Science Journalist (1986)

85.33 Astrophysicists and nuclear physicists together have concluded that the universe started out with only hydrogen and its isotope deuterium.

Dietrick E. Thomsen, American Science Journalist (1986)

85.34 Specialists . . . are slowly coming to the realization that the universe is biased and leans to the left. . . . Many scientists have come to believe that this odd state of affairs has something to do with the weak nuclear force. It seems that the weak force tends to impart a left-handed spin to electrons, and this effect may bias some kinds of molecular synthesis to the left. . . . But scientific speculation of this ilk leads to a deeper question. Was it purely a matter of chance that left-handedness became the preferred direction in our universe, or is there some reason behind it? Did the sinister bent of existence that scientists have observed stem from a roll of the dice, or is God a semiambidextrous southpaw?

Malcolm W. Browne, American Science Journalist (1986)

85.35 Astrophysicist David Schramm of the University of Chicago says that people may deceive themselves if they imagine the Big Bang as a kind of explosion. He prefers to use the analogy of raisin-bread dough. As the dough rises, it expands everywhere fairly evenly. As a result, the raisins are carried farther and farther from each other, even though they do not move with respect to the dough right around them. Galaxies, like the raisins, are carried farther and farther from each other as the space between them (like the dough) expands, but they do not necessarily move with respect to the space right around them.

Dietrick E. Thomsen, American Science Journalist (1987)

86. Zoology

If there is a just God, how humanity would writhe in its attempt to justify its treatment of animals.

ISAAC ASIMOV

86.1 And of every living thing of all flesh, you shall bring two of every sort into the ark, to keep them alive with you; they shall be male and female. Of the birds according to their kinds, and of the animals according to their kinds, of every creeping thing of the ground according to its kind.
The Bible **(circa 725 B.C.)**

86.2 Like sheep issuing from the fold by one, pairs, or three at once, while the others stand still, bending their eyes and noses to the ground, and what the foremost does that do the others, crowding behind her if she stops, simple and quiet, while they know not the cause.
Dante Alighieri, Italian Poet (1265–1321)

86.3 It is the wisdom of the crocodiles, that shed tears when they devour.
Francis Bacon, English Philosopher/Essayist/Statesman (1561–1626)

86.4 When it drew to night they were much perplexed, for they could find neither harbor nor meat, but, in frost and snow were forced to make the earth their bed and the element their covering. And another thing did very much terrify them; they heard, as they thought, two lions roaring exceedingly for a long time together, and a third, that they thought was very near them. So, not knowing what to do, they resolved to climb up into a tree as their safest refuge, though that would prove an intolerable cold lodging; so they stood at the tree's root, that when the lions came they might take their opportunity of climbing up. The bitch they were fain to hold by the neck, for she would have been gone to lion; but it pleased God so to dispose, that the wild beasts came out.

Anonymous (1620)

86.5 The surface of Animals is also covered with other Animals, which are in the same manner the Basis of other Animals that live upon it.

Joseph Addison, English Essayist/Poet/Politician (1672–1719)

86.6 Beasts have not the high advantages which we possess; but they have some which we have not. They have not our hopes, but then they have not our fears; they are subject like us to death, but it is without being aware of it; most of them are better able to preserve themselves than we are, and make a less bad use of their passions.

Baron de la Brède et de Montesquieu, French Philosopher/Author (1689–1755)

86.7 It is, I find, in zoology as it is in botany: all nature is so full that that district produces the greatest variety which is the most examined.

Gilbert White, English Naturalist/Ecologist (1720–1793)

86.8 The beavers are the philosophers of the animals; the gentlest, the most humble, the most harmless. Yet brutal Man kills them. I was once a witness to the destruction of one of their associated confederacies. I saw many of them shed tears, and I wept also; nor am I ashamed to confess.

Hector St. John de Crevecoeur, American Farmer/French Consul to New York (1735–1813)

86.9 Animal life is so infinitely abundant, and in forms so various, and so novel to European eyes, that it is absolutely necessary to divest oneself of all the petty terrors which the crawling, creeping, hopping, and buzzing tribes can inspire, before taking an American summer ramble. It is, I conceive, quite impossible for any description to convey an idea of the sounds which assail the ears from the time the short twilight begins, until the rising sun scatters the rear of darkness, and sends the winking choristers to rest.

Frances Trollope, English Writer (1780–1863)

86.10 Here is the distinct trail of a fox stretching [a] quarter of a mile across the pond. . . . The pond his journal, and last night's snow made a *tabula rasa* for him. I know which way a mind wended this morning, what horizon it faced, by the setting of these tracks; whether it moved slowly or rapidly, by the greater or less intervals and distinctness, for the swiftest step leaves yet a lasting trace.

Henry David Thoreau, American Writer/Naturalist (1817–1862)

86.11 Master said, God had given men reason, by which they could find out things for themselves; but He had given animals knowledge, which did not depend on reason, and which was much more prompt and perfect in its way, and by which they had often saved the lives of men.

Anna Sewell, British Writer (1820–1878)

86.12 Living organisms are not ordinarily given the proper amount of credit for their part in forming the character of the earth's surface.

C.C. Furnas, American Chemical and Aeronautical Engineer (1939)

86.13 The discoveries of Darwin, himself a magnificent field naturalist, had the remarkable effect of sending the whole zoological world flocking indoors, where they remained hard at work for fifty years or more, and whence they are now beginning to put forth cautious heads into the open air.

Charles Elton, English Ecologist (1960)

86.14 Mammals in general seem to live, at best, as long as it takes their hearts to count a billion. To this general rule, man himself is the most astonishing exception.

Isaac Asimov, American Biochemist/Author (1965)

86.15 A female ringtailed lemur (*Lemur catta*) edges toward another female holding a day-old infant in her arms. The first female sits down beside the mother, then suddenly grabs the baby and runs off. Possessing the infant raises this female's status within her troop and lowers that of the real mother, once the group's dominant female.

Jeffrey P. Cohn, American Science Journalist (1985)

86.16 Among those who study animal behavior, female competition has taken center stage during the last three years. In journals and at conferences, there have been reams of papers presented on the strategies of female animals—most notably, on how they go about sabotaging each other's reproductive cycles: aborting, starving, and even arranging the murder of each other's offspring. The more familiar offenders include macaques, chimpanzees, gorillas, gerbils, rabbits, wolves, and lions. . . .

In the last two decades field researchers have discovered that despite all our sensational murders and wars, human beings end up killing a smaller proportion of their fellows than do many other vertebrate species.

Duncan Maxwell Anderson, American Writer (1986)

86.17 The heaviest possible animal able to walk on four legs should have weighed no more than 100 tons.

This surprising conclusion will no doubt be tested by further discoveries of large fossil dinosaurs. But we must bear in mind that, biomechanically, bumblebees should not be able to fly, and dolphins should not be able to swim as fast as they do.

Mike Benton, Science Journalist (1986)

Blue Cliff Editions has conducted an exhaustive search to locate the publishers or copyright owners of the quotations in this book. However, in the event that we have inadvertently published previously copyrighted material without the proper acknowledgment, we advise the copyright owner to contact us so that we may give appropriate credit in future editions.

"A Response to Wang Ssu-Yuan's Poem on the Moon" by Shen Yueh, translated by Richard B. Mather from the book *Sunflower Splendor* by Wu-chi Liu and Irving Yucheng Lo. Copyright © 1975 by Wu-chi Liu and Irving Yucheng Lo. Reprinted by permission of Doubleday, a division of Bantam, Doubleday, Dell Publishing Group, Inc.

From *Collected Poems of Alfred Noyes*, an excerpt from "Watchers of the Sky" and from "Linnaeus," copyright © 1930, 1934, 1936, 1937, 1941, 1942, 1943, 1947 by Alfred Noyes. Reprinted by permission of Harper & Row, Publishers Inc.

Poem # 396 from *Bolts of Melody: New Poems of Emily Dickinson*, edited by Mabel Loomis Todd and Millicent Todd Bingham, copyright © 1945 by the Trustees of Amherst College. Reprinted by permission of Harper & Row, Publishers Inc.

Inscription by Darius I, excerpted from *Gods, Graves and Scholars*, Second Revised Edition, by C.W. Ceram, translated by Edward B. Garside and Sophie Wilkins. Copyright © 1951, 1967 by Alfred A. Knopf, Inc. Reprinted by permission of the publisher.

"Song of the Open Road" by Ogden Nash from *Verses from 1929 on* by Ogden Nash, copyright © 1932 by Ogden Nash. By permission of Little, Brown and Company.

Five haiku from *An Introduction to Haiku* edited by Harold G. Henderson. Copyright © 1958 by Harold G. Henderson. Reprinted by permission of Doubleday Publishing Group.

Excerpts from *Mayo: The Story of My Family and My Career* by Charles W. Mayo, M.D. Copyright © 1968 by Doubleday and Company. Reprinted by permission of the publisher.

Poem from *Hesiod: Theogony, Works and Days* by Apostolos N. Athanassakis, the Johns Hopkins University Press, Baltimore/London, copyright © 1983. Reprinted by permission.

Excerpt from *Little Willie*, lovingly collected by Dorothy Rickard. Illustrated by Robert Day. New York: Doubleday, 1953. Reprinted by permission.

The first ten lines of "The DNA Molecule" by May Swenson is used by permission of the author, copyright © 1968 by May Swenson.

Index

A

B

D

E

F

G

H

K

L

M

N

O

P

Q

R

S

T

U

V

W

X

Y

Z